SRA

Connecting Math Concepts

Level A Teacher's Guide

COMPREHENSIVE EDITION

A DIRECT INSTRUCTION PROGRAM

Mc Graw Hill Education

Bothell, WA • Chicago, IL • Columbus, OH • New York, NY

Acknowledgments

The authors are grateful to the following people for their input in the field-testing and preparation of *SRA Connecting Math Concepts: Comprehensive Edition Level A:*

Amilcar Cifuentes
Dorothy Glewwe
Crystal Hall
Joanna Jachowicz
Debbi Kleppen
Margie Mayo
Richelle Owen
Mary Rosenbaum
Jason Yanok

mheducation.com/prek-12

Send all inquiries to:
McGraw-Hill Education
4400 Easton Commons
Columbus, OH 43219

ISBN: 978-0-07-655572-7
MHID: 0-07-655572-0

Printed in the United States of America.

19 LKV 25 24 23

Contents

Program Overview
The *Connecting Math Concepts: Comprehensive Edition* Series

Connecting Math Concepts: Comprehensive Edition is a six-level series that will accelerate the math learning performance of children in grades K through 5. Levels A through F are suitable for regular-education students in Kindergarten through fifth grade. The series is also highly effective for at-risk students.

Connecting Math Concepts: Comprehensive Edition is based on the belief that understanding mathematics requires making connections—

- Among related topics in mathematics
- Between procedures and knowledge

Connecting Math Concepts does more than expose students to connections. It helps them structure their thinking by stressing understanding and by the way concepts are introduced and woven together throughout the program. Once something is introduced, it never goes away. It often becomes a component part of an operation that has several steps.

The organization of *Connecting Math Concepts: Comprehensive Edition* is powerful because lessons have been designed to:

- Teach explicit strategies that all students can learn and apply
- Introduce concepts at a reasonable rate, so all students make steady progress
- Help students make connections between important concepts and key ideas
- Maximize instructional time, so that all students have an opportunity to learn what is taught
- Provide the practice needed to achieve mastery and understanding
- Meet the math standards specified in the Common Core State Standards

The program's Direct Instruction design permits significant acceleration of student performance for both high performers and students who are at risk. The instructional sequences are the same for all students, but the rate at which students proceed through each level may be adjusted according to student performance. Higher performers proceed through the levels faster. Lower performers receive more practice.

Benchmark in-program Mastery Tests provide information about how well students are mastering what has been introduced most recently. Students' daily performance and test performance disclose whether they need more practice or whether they are mastering the material on the current schedule of lesson introduction.

Scripted lessons provide a rigorous approach to instruction that has been shaped through extensive field testing and classroom observation. The teacher engages students, adapting the presentation to the responses of each student. The teacher individualizes instruction to accommodate children making different mistakes and requiring different amounts of practice to learn the material on schedule. The program enables the teacher to teach children at a faster rate and with greater understanding than they probably ever achieved before.

Introduction to 2012 CMC Level A

- 2012 *CMC Level A* has a substantial amount of oral work.
- *CMC Level A* includes not only the basics but also concepts often not taught until third grade.
- Component skills and operations are keyed to kindergarten Common Core State Standards for Mathematics.

Level A is designed for students in grades K and above who can't identify numbers and who may not be able to count, but who have the basic language skills needed for them to follow the teacher's direction. The program is designed to greatly accelerate the math learning of both at-risk children and higher performers.

Children who complete *CMC Level A* will have a strong understanding of the fundamental logic of math as it applies to addition, subtraction, word problems, and the relationship of base-10 numerals to addition. For example, children learn that for any two-digit number, such as 62, there is an addition fact that is based on the digits of the number: $60 + 2 = 62$.

Program Information

The following summary table lists facts about
2012 *CMC Level A*.

Children who are appropriately placed in Level A	Pass Placement Test (p. 22)
How children are grouped	Small, homogeneous groups of 10 or fewer children (if possible)
Number of lessons	• 120 regular lessons • 12 test lessons (Mastery Tests) • 11 optional parallel lessons (1P, 3P, 5P, 6P, 9P, 11P, 15P, 17P, 22P, 41P, 59P)
Schedule	• 30–40 minutes for structured work • Additional 5–15 minutes for children's independent work • 5 periods per week

Teacher Material	• Teacher's Guide
	• Presentation Book 1:
	    Lessons 1–40, Mastery Tests 1–4, Answer Keys 1–40
	    9 Parallel Lessons
	• Presentation Book 2:
	    Lessons 41–80, Mastery Tests 5–8, Answer Keys 41–80
	    2 Parallel Lessons
	    Cumulative Test 1 (Lessons 1–60)
	• Presentation Book 3:
	    Lessons 81–120, Mastery Tests 9–12, Answer Keys 81–120
	    No Parallel Lessons
	    Cumulative Test 2 (Lessons 1–120)
	• Board Displays

Student Material	• Workbook 1: Lessons 1–60
	• Workbook 2: Lessons 61–120
	• Student Assessment Book: Mastery Tests 2–12,
	    Remedies worksheets for Mastery Tests 2–12
	    Cumulative Tests 1 and 2

In-Program Tests	12 ten-lesson Mastery Tests • Administration and remedies are specified in the Teacher Presentation Books. • Test sheets and Remedies worksheets are in the Student Assessment Book. • Reproducible Group Summary Sheets are in Appendix B of this guide.
Optional Cumulative Tests	2 Cumulative Tests • Administration is specified in Teacher Presentation Books 2 and 3. • Test sheets are in the Student Assessment Book. • Reproducible Group Summary Sheets are in Appendix A of this guide.
Additional Teacher/Student Material	• Math Fact Worksheets (Online Blackline Masters via ConnectED) • Access to CMC content online via ConnectED • *SRA 2Inform* available on ConnectED for online progress monitoring

TEACHER MATERIAL

The teacher material consists of:

1. **The Teacher's Guide.** This guide explains the program and how to teach it properly. The Scope and Sequence chart on pages 8–9 shows the various tracks, skills, or strands that are taught; indicates the starting lesson for each track/strand; and shows the lesson range. This guide calls attention to potential problems and provides information about how to present exercises and how to correct specific mistakes the children may make. The guide is designed to be used to help you teach more effectively.

2. **Three Teacher Presentation Books.** These books specify each exercise in the lessons and tests to be presented to the children. The exercises provide scripts that indicate what you are to say, what you are to do, the responses children are to make, and correction procedures for common errors. (See Teaching Effectively, **Using the Teacher Presentation Scripts,** for details about using the scripts.) The answers to all of the problems, activities, and tests appear in Teacher Presentation Books 1, 2, and 3 to assist you in checking the children's classwork, independent work, and tests.

   Book 1: Lessons 1–40; optional Parallel Lessons, 1P, 3P, 5P, 6P, 9P, 11P, 15P, 17P, 22P

   Mastery Tests 1–4

   Mastery Test Remedy specification

   Workbook and Test Answer Keys

   Book 2: Lessons 41–80; optional Parallel Lessons, 41P and 59P

   Mastery Tests 5–8; optional Cumulative Test 1

   Mastery Test Remedy specification

   Workbook and Test Answer Keys

   Book 3: Lessons 81–120 (no Parallel Lessons)

   Mastery Tests 9–12; optional Cumulative Test 2

   Mastery Test Remedy specification

   Workbook and Test Answer Keys

3. **Board Displays.** The teacher materials include online Board Displays which show all the displays you are to show or are to "write on the board." You may use the online displays as an alternative for displaying the book and for actually writing on the board. This component is flexible and can be utilized in different ways to support the instruction— via a computer hooked up to a projector, to a television, or to any interactive white board. The Board Displays are available online on ConnectED. You can navigate through the displays with a touch of the finger if you have an interactive white board or with the click of a button from a mouse (wired or wireless) or a remote control.

4. **ConnectED.** On McGraw-Hill/SEG's ConnectED platform you can plan and review *CMC* lessons and see correlations to Common Core State Standards for Mathematics. Access the following *CMC* materials from anywhere you have an Internet connection: PDFs of the Presentation Books, an online planner, online printable Board Displays, Math Fact Worksheets BLMs, eBooks of the Teacher's Guides, and correlations. *CMC* on ConnectED also features a progress monitoring application, called *SRA 2Inform*, that stores student data and provides useful reports and graphs about student progress. Refer to the card you received with your teacher materials kit for more information about redeeming your access code good for one six-year teacher subscription and 10 student seat licenses, which provides the ability to roster students in *SRA 2Inform*.

STUDENT MATERIAL

The student materials include a set of two workbooks for each child and a Student Assessment Book. The workbooks contain writing activities, which the children do as part of the structured presentation of a lesson and as independent seatwork. The Student Assessment Book contains material for the Mastery Tests as well as test Remedies worksheet pages and optional cumulative test pages.

Workbook 1: Lessons 1–60; optional Parallel Lessons 1P, 3P, 5P, 6P, 9P, 11P, 15P, 17P, 19P, 22P, 41P, and 59P

Workbook 2: Lessons 61–120 (no Parallel Lessons)

Student Assessment Book: Mastery Tests 2–12 (Test 1 is oral), Cumulative Test 1 and 2, and test Remedies worksheets for Mastery Tests

WHAT'S NEW IN 2012 *CMC LEVEL A*

Most instructional strategies are the same as those of the earlier *CMC* editions; however, the procedures for teaching these strategies have been greatly modified to address problems teachers had teaching the content of the previous editions to at-risk children. The 2012 edition of *Level A* has also been revised substantially two times on the basis of field testing.

1. The 2012 edition provides far more oral work, which is presented through "hot series" of tasks. The series are designed so that children respond to ten or more related questions or directions per minute; therefore, these series present a great deal of information about an operation or discrimination in a short period of time. Each lesson provides at least two times the amount of practice provided by comparable lessons in the earlier editions.

2. The content is revised so that children learn not only the basics but also the higher-order concepts that children sometimes do not master by the third grade. The result is that kindergarten children who complete *Level A* are able to solve problems of the following types:

   - Two-digit addition problems, including those that are traditionally solved by carrying, for example:

$$
\begin{array}{r}
37 \\
+\ 24 \\
\hline
\end{array}
$$

   - Two-digit subtraction problems that require no borrowing or renaming:

$$
\begin{array}{r}
47 \\
-\ 16 \\
\hline
\end{array}
$$

(The procedures for solving these problems are introduced in the **Practice Tasks** section, pages 38–43, and explained further in the **Tracks** section, which follows the practice tasks.)

3. *CMC A* teaches the component skills and operations required to provide a solid foundation. The program now includes the topics of money, geometry and measurement and data classification and organization. The program addresses all standards specified in the Common Core State Standards for Kindergarten Mathematics. (pages 10–11 and 120–134.)

4. *CMC Level A* offers numerous support/ enhancements, including technology components, for teachers and students. These enhancements include displays in the Teacher Presentation Books, online printable Board Displays, Workbook and Answer Key pages reduced in the Teacher Presentation Book, a Student Assessment Book with all program assessments in one location, *SRA 2Inform* for online progress monitoring, and the ability to plan and review lessons online via ConnectED.

Preparing to Teach the Program

TIME REQUIREMENTS FOR CONNECTING MATH CONCEPTS

A 45-minute period should be scheduled on every available school day. Schedule an additional 30 minutes a day for each additional group. Ideally children should be divided into small, homogeneous groups for instruction (see Teaching Effectively, **Organization**), unless there is no alternative to teaching the entire class at once. The time required for the various activities is:

- Teacher instruction of groups: 30–40 minutes a day for each group
- Children's independent work: 5–15 minutes for each group

Students who are appropriately placed in the program should complete one lesson per period.

PRESENTATION FORMATS

In *CMC Level A*, there are four basic presentation formats:

- Oral
- On the board (or Board Displays)
- From a display page (or Board Displays)
- In the Workbook

The program presents oral tasks, tasks that display material for children to observe, and workbook tasks.

The wording for presenting all tasks is in the Teacher Presentation Books.

Oral

The oral formats direct children to produce oral responses. For example, "Everybody count to ten. Get ready…" Children respond orally.

On the Board

This format involves displays that children view and respond to. For instance, the exercise directs the teacher to write 7 − 4 on the board.

(Display:) W

7 − 4 =

- (Point to 7 − 4.) Read this problem. (Signal.) *7 take away 4.*

W indicates that this is a display the teacher can write on the board. The script indicates that the teacher is to say, "Read this problem." The script that follows, which is not shown here, indicates the children's responses, what the teacher does and says next, and how the teacher changes the display.

From a Display: Presentation Book

The teacher displays the material children will respond to. This material is in the Teacher Presentation Book. The teacher holds the book so that children can see the display. The teacher follows the script by pointing to parts of the display and presenting questions or directions. For instance, the teacher displays a page that shows 2-digit numbers. The teacher points to 17 and says, "What's this number?" Children respond.

Board Displays

The Board Displays provide an alternative for presenting displays. All On-the-Board displays and all the displays in the Teacher Presentation Books are available online. To use the displays, you would project them onto a screen, the wall, or an interactive white board. Move between displays using any mouse or remote control. The Board Displays are accessed online from ConnectEd.

In the Workbook

Workbook formats direct children to perform tasks in their Workbook. These are usually tasks that have been practiced in earlier oral and display formats; however, skills like symbol writing are introduced for the first time in the Workbook. A structured teacher presentation accompanies one or more problems on each worksheet. The script for Workbook exercises indicates what the teacher says and children's responses.

Independent Work

In addition to the teacher-led activities, each worksheet contains problems and activities for 5–15 minutes of independent work, which is scheduled daily at any convenient time.

Children should not receive help when they are working independently on the worksheets, except early in the program when some children need help with writing. Giving children the answers to problems encourages them to become dependent on you. You will receive muddled feedback about the effectiveness of your teaching if you help the children when they work independently. Let them know that they are to work alone. Their mistakes will tell you whether additional work is needed on specific skills.

Collect the Workbooks each day and mark the mistakes. Answer Keys appear in the Presentation Books. Note common mistakes that the children make and plan a time for practice on the weak skills. Require children to correct all errors in their Workbooks before taking them home.

STUDENT ASSESSMENT BOOKS

The student pages for the in-program Mastery Tests and optional Cumulative Tests appear in the Student Assessment Book. (The teacher test material appears in the Presentation Books.) Remedy worksheets for the Mastery Tests also appear in the Student Assessment Book.

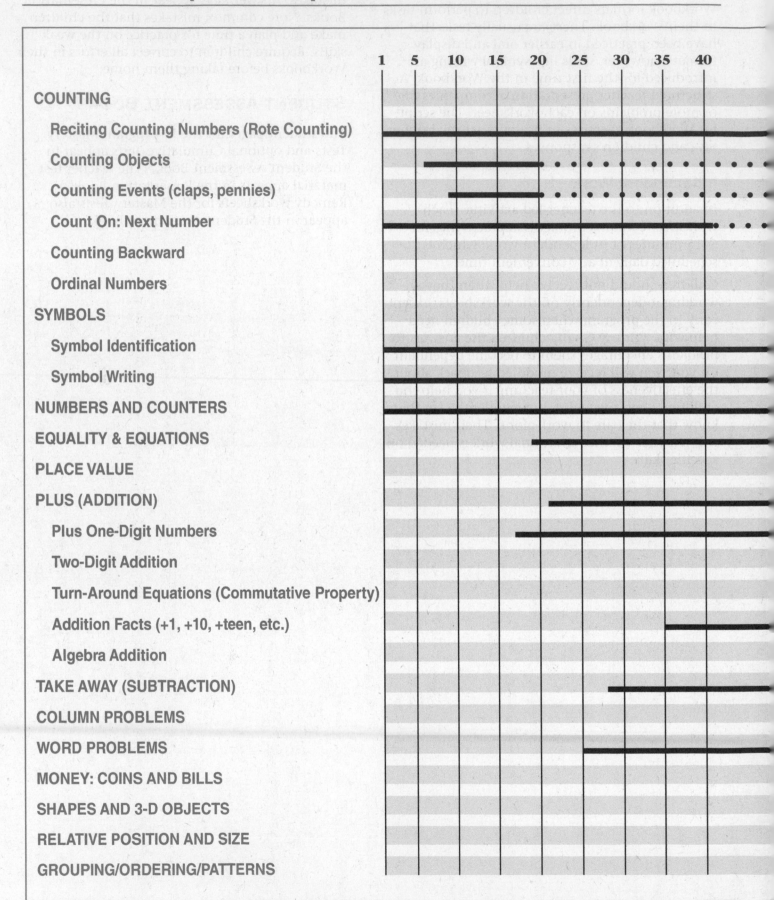

	1	5	10	15	20	25	30	35	40
COUNTING									
Reciting Counting Numbers (Rote Counting)									
Counting Objects									
Counting Events (claps, pennies)									
Count On: Next Number									
Counting Backward									
Ordinal Numbers									
SYMBOLS									
Symbol Identification									
Symbol Writing									
NUMBERS AND COUNTERS									
EQUALITY & EQUATIONS									
PLACE VALUE									
PLUS (ADDITION)									
Plus One-Digit Numbers									
Two-Digit Addition									
Turn-Around Equations (Commutative Property)									
Addition Facts (+1, +10, +teen, etc.)									
Algebra Addition									
TAKE AWAY (SUBTRACTION)									
COLUMN PROBLEMS									
WORD PROBLEMS									
MONEY: COINS AND BILLS									
SHAPES AND 3-D OBJECTS									
RELATIVE POSITION AND SIZE									
GROUPING/ORDERING/PATTERNS									

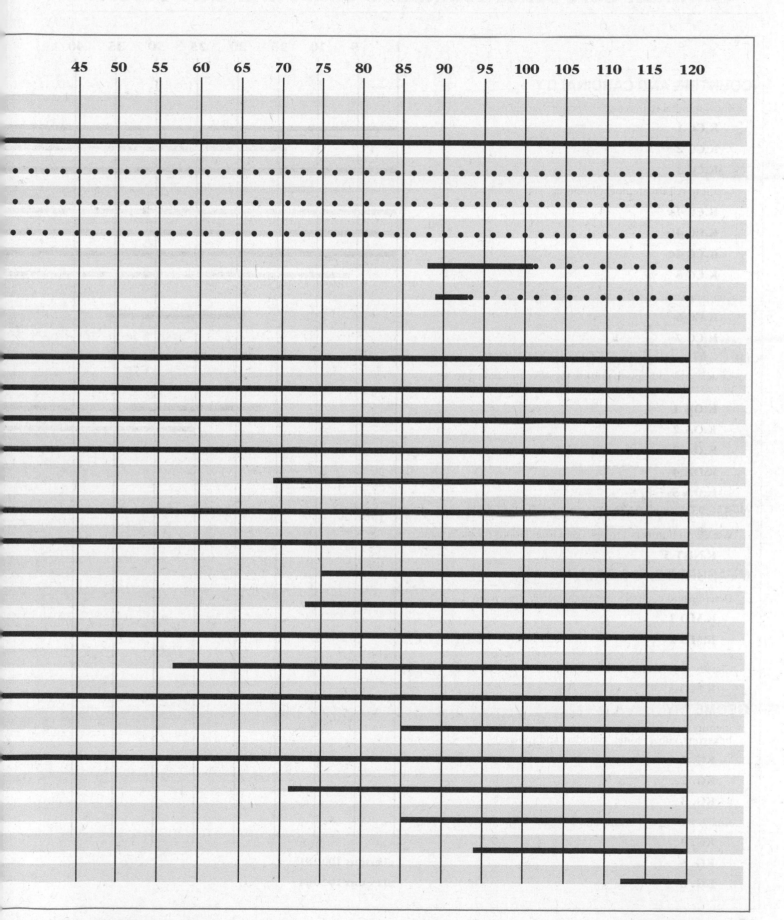

Common Core State Standards Chart and CMC Level A

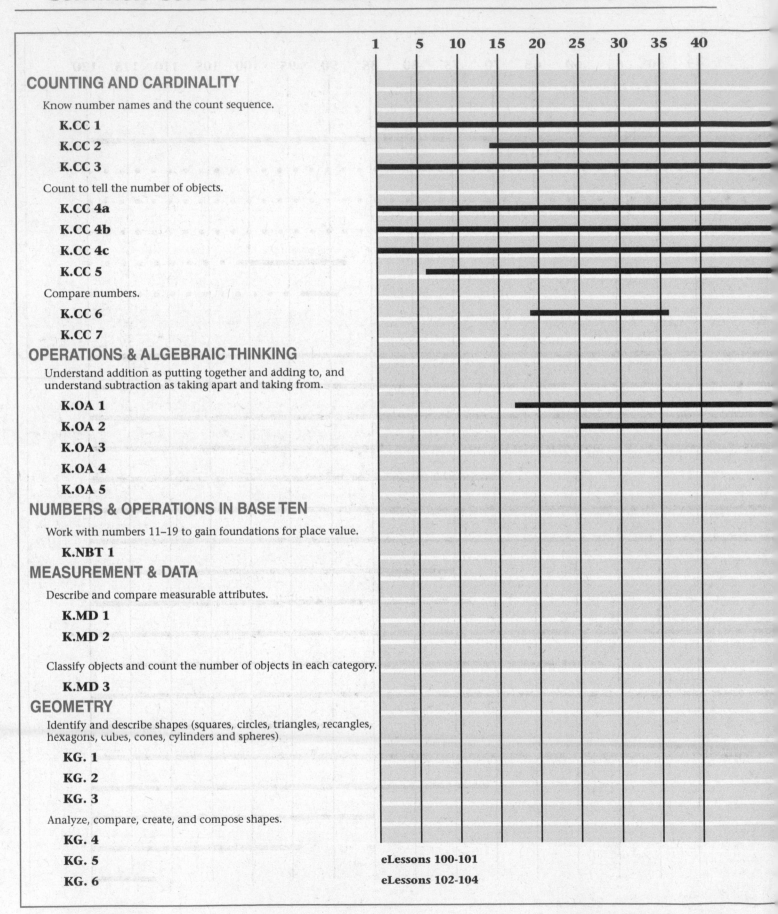

	1	5	10	15	20	25	30	35	40

COUNTING AND CARDINALITY

Know number names and the count sequence.

K.CC 1

K.CC 2

K.CC 3

Count to tell the number of objects.

K.CC 4a

K.CC 4b

K.CC 4c

K.CC 5

Compare numbers.

K.CC 6

K.CC 7

OPERATIONS & ALGEBRAIC THINKING

Understand addition as putting together and adding to, and understand subtraction as taking apart and taking from.

K.OA 1

K.OA 2

K.OA 3

K.OA 4

K.OA 5

NUMBERS & OPERATIONS IN BASE TEN

Work with numbers 11–19 to gain foundations for place value.

K.NBT 1

MEASUREMENT & DATA

Describe and compare measurable attributes.

K.MD 1

K.MD 2

Classify objects and count the number of objects in each category.

K.MD 3

GEOMETRY

Identify and describe shapes (squares, circles, triangles, recangles, hexagons, cubes, cones, cylinders and spheres)

KG. 1

KG. 2

KG. 3

Analyze, compare, create, and compose shapes.

KG. 4

KG. 5   eLessons 100-101

KG. 6   eLessons 102-104

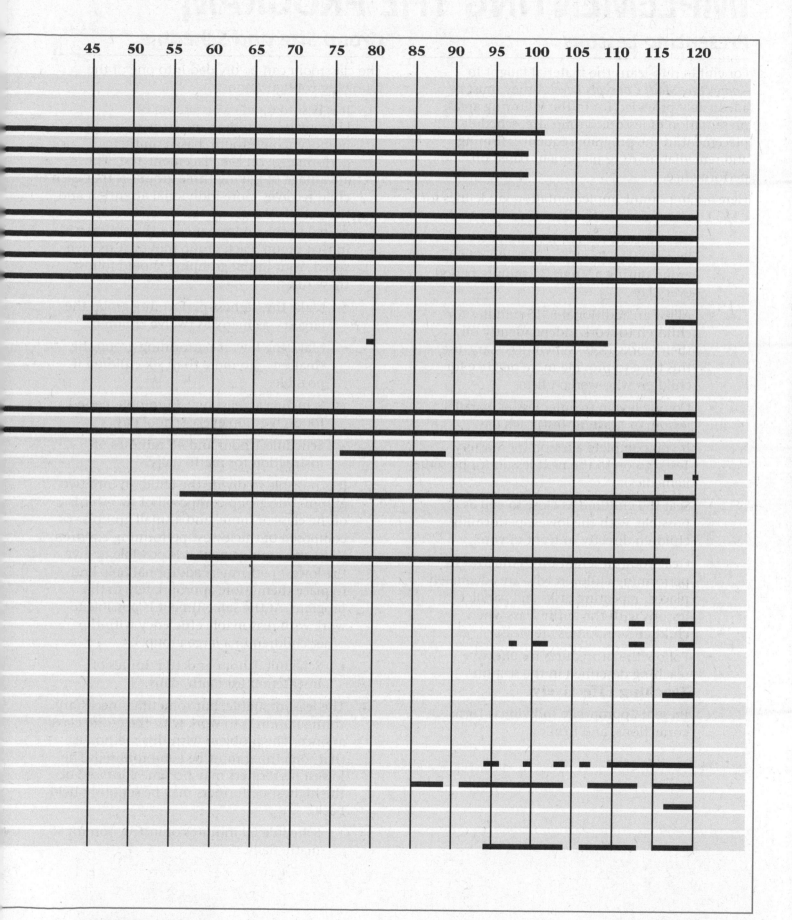

IMPLEMENTING THE PROGRAM

Presenting Lessons

For children to learn the material taught in *Connecting Math Concepts Level A* they must be adequately provided for in the following areas: presentation of lessons, group size, schedule, placement in the program sequence, firming, and remediation or re-placement based on performance.

Here are important implementation guidelines for *CMC Level A:*

- Present each regular lesson in the Presentation Book to all children in a group during a single 35-minute period, everyday.

- Allow an additional 5–15 minutes for children to work independently on their worksheets. When necessary, use this time to provide more practice for children who weren't firm.

- Do one lesson (regular lesson, parallel lesson, or Mastery Test) each day.

- If you complete a lesson (or Mastery Test), go on to the next lesson (or present Remedies).

- Seat the children as close to you as possible, with the lowest-performing children directly in front of you.

- Gear your presentation to the lowest-performing children who are adequately placed, repeating tasks and parallel lessons with the entire class when children's responses are weak.

- Follow the procedures for effective teaching described in the section **Teaching Effectively.**

- Present appropriate Individual Turns, corrections, and firming.

Group Size and Schedule

The classroom can be divided into one of the following configurations:

1. It is best to divide a classroom of kindergarten children into three homogeneous groups, based on their performance on the Placement Test. The advantage of having three groups is that each child receives more individual attention and the pace of the group can be more closely geared to the performance of the children in that group. Each group moves at its own speed. Your initial grouping should follow these rules:

   - Make the highest-performing group the largest—as many as twelve children.

   - Make the lowest-performing group the smallest—no more than six children if possible.

   - Schedule at least one 35-minute period for each group every school day.

   - Schedule 1 hour and 45 minutes of instruction for math, daily.

2. It is possible to divide the children into two groups: one group composed of two-thirds of the class or more, and a small group composed of the lowest-performing children. With this arrangement it is possible to give the lowest performers additional help and to place them more appropriately in the program. At the same time it is possible to work with reasonable efficiency with the other children in a large group.

   - Schedule 1 hour and 10 minutes of instruction for math, daily.

3. The least desirable, but sometimes necessary, configuration is to work with the entire class at once. The problem with this method is that something must be compromised: The lowest performers may become confused or the highest performers may be seriously held back.

   - Schedule 50 minutes of instruction for math, daily.

Placement

Performance data from three sources can help determine an appropriate placement for children in the *CMC* sequence. This data comes from formal assessments, in-class performance, and individual turns.

CMC LEVEL A PLACEMENT TEST

A complete description about placement testing for *CMC Level A* and the full-sized Placement Test begins on page 22 of this guide.

Before you begin teaching the program, administer the test to each child individually. You should be able to test the children in a *CMC Level A* classroom within one hour on the first day of school. Instruction should begin on the second day.

Note: Children who do not pass the Placement Test need to receive language instruction before beginning *CMC Level A. Language for Learning* is designed to teach the skills these children need.

The Placement Test is the primary tool for determining if children can begin instruction in the *CMC Level A* program.

MASTERY TESTS

Mastery Tests provide information about how well prepared each child is to proceed in the *CMC Level A* program. Children's performance on Mastery Tests can also be used to place children on an appropriate lesson in the *CMC* sequence.

There are 12 in-program Mastery Tests in *CMC Level A*. The tests appear every 10 lessons and evaluate children's performance on the critical skills that have been introduced, practiced, and applied, but not tested. Each Mastery Test provides information about how to score children's responses, criterion for passing each part, and remedies for teaching the skills for parts children fail. A complete discussion about presenting, grading, and providing remedies for Mastery Tests begins on page 14.

PLACEMENT OF CHILDREN MID-SEQUENCE

If a child transfers into your classroom very early in the program, administer the Placement Test to determine whether the child qualifies to enter the program. If so, and if you have more than one instructional group (three is optimal), place the child in an appropriate group based on the Placement Test results. Observe the child closely after this initial placement to be sure the placement is appropriate (see **Adjusting Placement** in this section on page 14).

Children transferring into the program later in the year are more difficult to place initially. Some children may require special tutoring before they can enter even the lowest-performing group in a classroom.

Initially place the child in the lowest group. If the child is performing well on group tasks and individual tests (individual turns) and mastery-test performance, move the child to the next highest group. Observe the child's performance closely until the next Mastery Test (about two weeks) and then evaluate the child again using in-class and the mastery-test performance. If the child's performance is a little weak at this point, don't be too hasty to move the child back to a lower group. Make allowances for the child's probable lack of familiarity with the teacher presentation vocabulary in the first weeks.

Some children have enough skills to be placed in the middle of the *CMC Level A* sequence. Here's a summary of the actions for placing these children:

1. Look at the Mastery Tests and choose a target test that approximates the child's skill level.
2. Go back three Mastery Tests and present it.
3. If the child passes all of the parts, go to step 7.
4. If the child fails parts, present the remedies for those parts.
5. Retest the child on parts failed.
6. Evaluate the retest
   a. If the child passes all parts of retest, go to step 7.
   b. If the child fails one or two parts of retest, place the child on the lesson which appears ten lessons before the failed Mastery Test.

c. If the child fails more than two parts, choose a target test that is less difficult and go back to step 1.

7. Present the next Mastery Test.

8. If the child fails parts, present remedies for those parts.

9. Retest the child on parts failed.

10. Evaluate the retest.

    a. If the child passes all parts of retest, go to step 7.

    b. If the child fails one or more parts of retest, place the child on the lesson which appears ten lessons before the failed Mastery Test.

ADJUSTING PLACEMENT

Your initial grouping of children is not permanent. A child may progress more or less rapidly than the group. Children should be moved to a more appropriate group when it becomes apparent that their performance in a particular group is no longer suitable.

How do you know if a child's placement is not suitable?

- The hundreds of group responses children are directed to produce each lesson provide a great deal of data about a child's mastery and fluency of the material. Children who regularly make mistakes, don't respond or respond slowly, or who consistently respond correctly and more quickly than the rest of the group are probably not placed adequately.

- The Individual Turns give information about the performance of children who respond. A child who consistently fails Individual Turns or who consistently passes them when other children in the group struggle is not placed adequately.

- The 12 in-program Mastery Tests provide you with even more detailed information about the performance of each child. If a child's performance is not consistent with the others in the group, you should consider moving the child. (The next section provides a complete discussion about Mastery Tests.)

Using Mastery Tests

There are 12 in-program Mastery Tests that are used to evaluate the children's performance on critical skills. The Mastery Tests appear in the Teacher Presentation Books at the end of every tenth lesson. The student material for the tests appears in the Student Assessment Book.

The Mastery Test instructions are very important. They tell you how to administer the test and what to do when children fail parts of the test. When remedies are needed, these are specified for presentation *after* the test and before the next lesson. The test lessons are usually shorter than regular lessons. A Mastery Test should never be skipped.

Test 1 is oral. Tests 2 through 12 require the use of the Student Assessment Book. This is clearly indicated in the test instructions. The last part of each test is individually administered. Use half the period to present the individually administered section. Try to complete this part of the test before beginning the next lesson. Begin presenting the remedies for the group-administered section or, if there are no remedies to present, finish testing individuals and then go to the next lesson.

The primary purpose of the tests is to provide you with information about how well prepared each student is to proceed through the program. The in-program tests permit you to assess how well each student is mastering the program content.

Sometimes children copy from their neighbors. A good method to prevent copying is to spread children out during the test if it's physically possible to do so. Discrepancies in the test performance and daily performance of some children pinpoint which children may be copying.

To avoid copying, reassign seating so that the children who tend to copy are either separated from those who know the answers or are seated near the front center of the room where it is easier for you to monitor them.

Following is the written portion of Mastery Test 4 and its Answer Key. Mastery Test 4 is scheduled after Lesson 40.

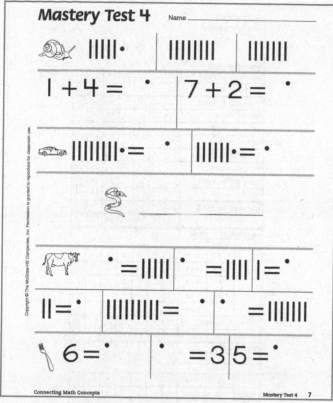

Mastery Test 4, Student Assessment Book

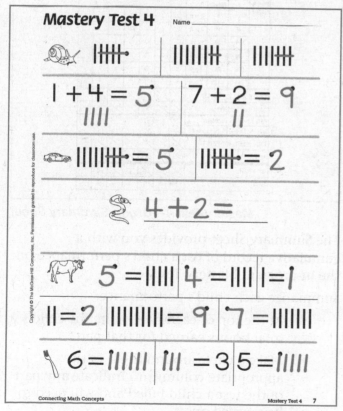

Mastery Test 4, Answer Key

Connecting Math Concepts

Parts 1–6 are in the Student Assessment Book and are administered to the group. The Answer Key is in the back of the Teacher Presentation Book.

Parts 7–10, below, are in the Teacher Presentation Book and are administered to children individually.

$$8 \quad - \quad 0 \quad 14 \quad 17$$

Individually Administered Section

Part 7: (15 points possible)

a. (Display page and point to 14 and 17.) These are teen numbers.
- (Point to **8**.) Touch each of these symbols and tell me what it is. *8, take away, zero, 14, 17.*

Part 8: (12 points possible)

a. Count from 18 to 28. Get 18 going and count. *Eighteeen, 19, 20, 21, 22, 23, 24, 25, 26, 27, 28.*
(If child wasn't able to count correctly the first time, present the task one more time. Do not present the task again if child counted correctly the first time.)

b. Count from 28 to 38. Get 28 going and count. *Twenty-eieieight, 29, 30, 31, 32, 33, 34, 35, 36, 37, 38.*
(If child wasn't able to count correctly the first time, present the task one more time. Do not present the task again if child counted correctly the first time.)

Part 9: (12 points possible)

I'll say numbers. You'll say the number, then tell me the next number.

a. {Non-scored item:} Listen: 4. What number? *4.*
- {Scorable item:} (Pause 2 seconds.) Next number? *5.*
b. {Non-scored item:} New number: 9. What number? *9.*
- {Scorable item:} (Pause 2.) Next number? *10.*
c. {Non-scored item:} New number: 19. What number? *19.*
- {Scorable item:} (Pause 2.) Next number? *20.*
d. {Non-scored item:} New number: 29. What number? *29.*
- {Scorable item:} (Pause 2.) Next number? *30.*
e. {Non-scored item:} New number: 25. What number? *25.*
- {Scorable item:} (Pause 2.) Next number? *26.*
f. {Non-scored item:} New number: 13. What number? *13.*
- {Scorable item:} (Pause 2.) Next number? *14.*

Part 10: (5 points possible)

Note: (Do not display page until step b.)

a. I'm going to tell you a story problem that pluses. You're going to tell me the symbols to write.
Listen: A teacher has 6 peanuts. Then the teacher gets 3 more peanuts.
- Listen again: A teacher has 6 peanuts.
{Scorable item, 1 point:} What symbol do I write for that part? *6.*
- Listen to the next part: Then the teacher gets 3 more peanuts.
{Scorable item, 3 points:} What symbols do I write for that part? *Plus 3.*
- We want to find out what 6 plus 3 equals.
{Scorable item, 1 point:} So what symbol do we write next? *Equals.*
b. (Display page and point to **6 + 3 =**.)
Here's the problem.

$$6 + 3 =$$

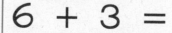

Mastery Test 4, Teacher Presentation

Each Mastery Test is worth 100 points. The Answer Key (which is in the back of the Teacher Presentation Book) provides the correct answers and shows the work for each item in the group-administered section. The Passing Criteria Table, which accompanies each test in the Teacher Presentation Book, gives the possible points for each part and the passing scores. Children fail a part of the test if they score fewer than the specified number of passing points.

A table accompanies each test showing the passing score for each part.

Here is the Passing Criteria table for Test 4.

Passing Criteria Table — Mastery Test 4											
Part	1	2	3	4	5	6	7	8	9	10	Total
Passing Points	6	8	6	5	15	9	15	8	10	5	
Possible Points	6	10	8	5	18	9	15	12	12	5	100

Scoring Mastery Tests

Use the criteria and key in the Teacher Presentation Book for marking each children's test. During the mastery test presentation, record the results on the **Mastery Test Group Summary Sheet** (which appears in Appendix B of this *Teacher's Guide*). The Group Summary Sheet can accommodate up to 15 children. Use duplicate copies to accommodate more children. The sample below summarizes the performance of 6 children on Mastery Test 4.

Mastery Test 4—Group Summary Sheet

The Summary Sheet provides you with a cumulative record of each child's performance on the in-program Mastery Tests.

Summarize each child's performance.

- At the top of each part, write the child's total points earned for that part.

- Make a circle around the R in the appropriate columns to indicate any part of the test a child failed and will need to be retested on.

- Circle Y for any part requiring a group remedy. Provide a group remedy for each part that has a failure rate of more than 25% (.25).

Test Remedies

The Teacher Presentation Book specifies remedies for each test. Any necessary remedies should be presented before the next lesson.

Here are the remedies for Mastery Test 4:

Part	Test Items	Remedy		Student Material Remedies Worksheet
		Lesson	Exercise	
1	Crossing Out Counters	38	7	Ⓐ on Worksheet
		39	7	Ⓑ on Worksheet
2	Addition	34	10	Ⓒ on Worksheet
		38	10 (steps h–i)	Ⓓ on Worksheet
3	Subtraction	39	9	Ⓔ on Worksheet
		40	6	Ⓕ on Worksheet
4	Addition	34	3 (steps a–c)	—
		35	11 (steps a–c)	Ⓖ on Worksheet
		38	10 (steps a–d)	Ⓗ on Worksheet
5	Equations	37	10	Ⓘ on Worksheet
		38	11 (steps a–d)	Ⓙ on Worksheet
6	Equals	31	8	Ⓚ on Worksheet
		33	9	Ⓛ on Worksheet
7	Identifying 8	32	5	—
	Identifying –	35	6	—
		38	3	—
	Identifying 0	29	6	—
	Identifying 14, 17	35	8	—
		38	5	—
8	Rote Counting (18–28)	32	1	—
		37	1	—
	Rote Counting (28–38)	36	3	—
		38	2	—
9	Next Number	33	6	—
		36	1	—
		37	3	—
10	Word Problems	32	7	—
		34	7	—
		36	6 (steps a–d)	—

If the same children consistently fail parts of the test, it may be possible to provide remedies for those children as the others do a manageable extension activity. If individual children are weak on a particular skill, they will have trouble later in the program when that skill becomes a component in a larger operation or more complex application.

If children consistently fail tests, they are probably not placed appropriately in the program.

On the completed Mastery Test 4 Group Summary Sheet, shown on the previous page, more than ¼ of the children failed parts 1, 3, and 10.

After you provide group remedies by reteaching the exercises specified as the remedy for those parts, you need to provide individual remedies for Amanda Adams, Henry Bowman, and Chan Won Lee because they failed additional parts (Amanda–Part 5; Henry–Parts 8 and 9; Chan–Part 2).

The goal of each remedy is to teach children well enough that they can work items of that type in the context presented in the *CMC Level A* program.

CHILDREN WHO FAIL THREE OR MORE PARTS OF THE MASTERY TEST

As a rule of thumb, if a child fails three or more parts of the test, the child is not placed properly in the program, which means that the child will continue to have problems with the material. The ideal remedy would be to place children at a lesson in which they could be successful on about 85% of the tasks on each exercise.

The children who failed the test must be brought to criterion if they are to continue in this group. Watch them closely during the next few days to be sure their placement in this group is still appropriate. When you present tasks children had trouble with on the Mastery Test, pay close attention to their performance and provide them with additional practice on those tasks if necessary.

MASTERY TEST RATIONALE

The in-program Mastery Test assessments are helpful to you in four ways:

- They give you feedback in the effectiveness of your teaching. For example, if you have a group that fails to reach criterion on one test after another, you should examine whether you need to make adjustments to your presentation. For example, make sure that the children are performing correctly on each task before you leave it. Be conscientious about the instructions "Repeat until firm" and "Repeat tasks that were not firm." Children must get enough repetition executing strategies so that they know exactly what to do to solve new problems. When errors occur, make sure you are using the error-correction procedures where specified in the Teacher Presentation Book.

- The tests serve as a backup for your daily evaluation of the children (individual tests) to help you make sure that individuals and groups are firm on each critical skill.

- The tests give you information that will help you regroup and place children appropriately. The *CMC Level A* Placement Test gives you information about how the children perform at the beginning of the year, but it does not tell you how fast individual children will progress. Some children who start with little skill can progress rapidly. Others start with more skill but progress more slowly. Anticipate needing to regroup children during the year.

- The tests may be used to place mid-year transfer children. (See **Placement of Children Mid-sequence**, earlier in this section on page 13).

Parallel Lessons

CMC Level A has eleven "P" lessons or parallel lessons. These are optional lessons that appear when the program introduces new material too quickly for some groups. P lessons don't introduce anything new—they simply present the same series of exercises that were presented in the corresponding regularly numbered lesson. For exercises in the P lessons that teach material newly introduced in the regularly numbered lesson, the same examples are usually used but the examples appear in a different order. For P-lesson exercises that review familiar material, the examples are often changed to other examples. P lessons provide teachers the option of slowing the introduction of new material and providing additional practice applying new and familiar concepts and information.

The judgment of whether you present a P lesson is based on the children's performance. Did they have trouble on three or four exercises in the preceding lesson, or did their performance seem quite solid at the end of the period? If they had trouble, present the P lesson that follows. If they didn't have trouble, skip the P lesson. If their performance seemed to be "borderline" or if you have doubts, present the P lesson. The additional practice will benefit children and promote fluency. Children should find the P lesson easy and the lesson should serve as good practice and review. After all, the goal of instruction is not to present material that children are able to respond to if they think hard about it, but to present enough practice for children to learn to respond "automatically."

The eleven P lessons follow the lessons listed below. Each P lesson has the lesson heading of Lesson 1P, Lesson 3P, and so forth. P lessons follow Lesson 1, 3, 5, 6, 9, 11, 15, 17, 19, 22, 41, and 59.

If children are not firm on Lesson 1, present the P lesson that follows (Lesson 1P).

Note that the frequency of P lessons is greatest in lesson range 1–10. This is where it is most important for children to be very firm on everything the program teaches. Also, the rate at which different children learn varies the most during the first lessons. So a good strategy for the first 20 lessons is to present the P lessons unless it is clear that children are **quite** firm on everything in the original lesson.

Cumulative Tests

There are two Cumulative Tests that accompany *CMC Level A*: Cumulative Test 1 assesses skills that are taught in the first half of the program (Lesson 1–60); Cumulative Test 2 assesses skills that are taught in the entire program (Lessons 1–120).

The Cumulative Tests have two potential uses: to assess children's mastery of the skills that they were taught in the first half or the entire *CMC Level A* program and to provide specific comprehensive examples of problem types covered.

Assessing Children's Mastery

The tests are well designed to serve as benchmark assessments for the first half and the end of the *CMC Level A* program. Children who have mastered the material should pass all parts of the Cumulative Tests and score 85% or better on the test (between 170 and 200 points). If most children are not close to achieving this level of performance, skills that children are deficient in should receive intensive and on-going firming. The teacher should consider repeating some earlier lessons as well.

Specific Comprehensive Examples

The Cumulative Tests are useful for illustrating to parents, community members, and administrators the skills children learn throughout the course of the program.

Cumulative Tests and Mastery Tests follow the same procedures for administration and scoring. A complete discussion about the Cumulative Tests, including specific examples, appears on ConnectED. Following are the answer keys for Cumulative Tests 1 and 2.

Cumulative Test 1

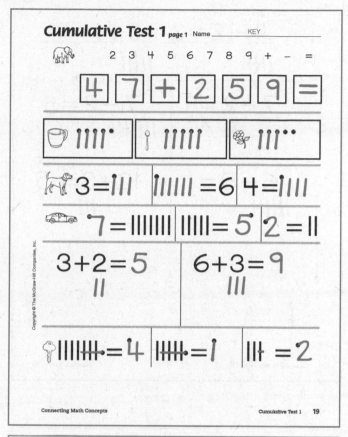

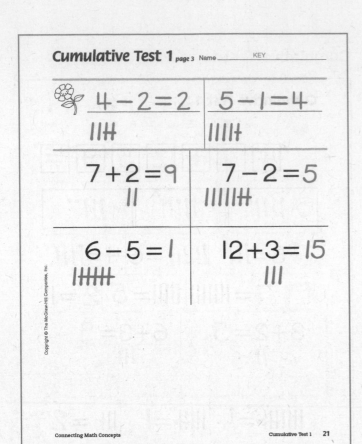

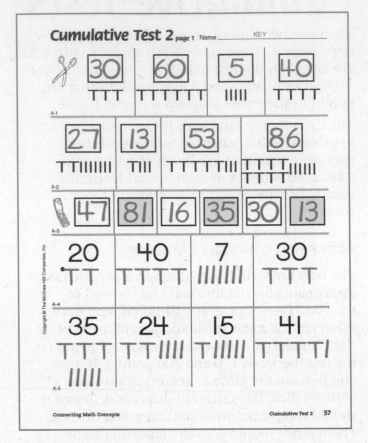

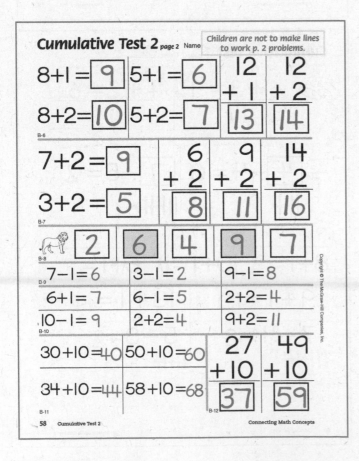

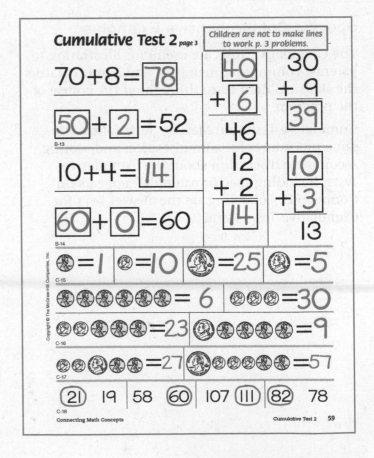

[$50] [$5] [$1] [$1] [$1] = 58

[$20] [$10] [$10] [$10] [$5] = 55

[$10] [$10] [$5] [$1] [$1] = 27

C-19

| 5 | 8 | 7 | 6 | 11 | 10 | 9 |

| 5 | 6 | 7 | 8 | 9 | 10 | 11 |

C-20

| $6 + \boxed{4} = 10$ | $12 + \boxed{3} = 15$ |
| IIII | III |

D-21

| $8 + 3 = \boxed{11}$ | $2 + \boxed{5} = 7$ |
| III | IIIII |

| $7 - 2 = \boxed{5}$ | $5 + \boxed{3} = 8$ |
| IIIIIII | III |

D-22

60   Cumulative Test 2   Connecting Math Concepts

| $36 + 40 = 76$ | $36 + 4 = 40$ |
| TTTT | IIII |

| $25 + 3 = 28$ | $25 + 30 = 55$ |
| III | TTT |

D-23

| $42 + 23 = 65$ | $37 + 43 = 80$ |
| TT III | TTTT III |

D-24

| $61 + 21 = 82$ | $58 + 34 = 92$ |
| TT I | TTT IIII |

D-25

| $57 - 23 = 34$ | $44 - 12 = 32$ |
| TTTT̶ IIIIIII | TTT̶ IIII |

E-26

| $65 - 34 = 31$ | $54 - 21 = 33$ |
| TTT̶ T̶ IIII | TTT̶ T̶ IIII |

E-27

| $36 - 22 = 14$ | $43 - 13 = 30$ |
| T̶ T̶ IIIIIII | TTT̶ III |

E-28

Connecting Math Concepts   Cumulative Test 2   61

| $\begin{array}{r}24\\+15\\\hline39\end{array}$ T IIIII | $\begin{array}{r}46\\-32\\\hline14\end{array}$ T̶ IIII IIIIIII |

| $34 - 31 = 3$ | $34 + 31 = 65$ |
| T̶ T̶ T̶ IIII | TTT I |

E-29

| $78 + 15 = \boxed{93}$ | $45 - 41 = \boxed{4}$ |
| T IIIII | T̶ T̶ T̶ T̶ IIIII |

E-30

| $87 - 69 = 18$ | $47 + 15 = 62$ |
| [blank] | $15 + 47 = 62$ |

| $64 + 25 = 89$ | $52 - 21 = 31$ |
| $25 + 64 = 89$ | [blank] |

F-31

62   Cumulative Test 2   Connecting Math Concepts

| 4 | 5 | 6 | 7 | 8 | 9 | 10 |

| $6 + 1 = 7$ | $7 - 1 = 6$ |
| $8 + 2 = 10$ | $10 - 2 = 8$ |

F-32

KMLMKMLPMLM

K = 2   **L** = 3   **M** = 5   **P** = 1

Letter	1	2	3	4	5	6
	P	K	L	[blank]	M	[blank]

G-33

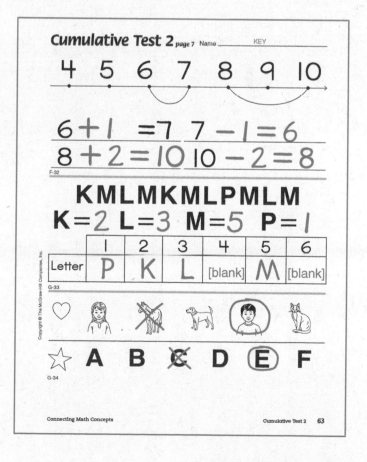

☆ **A** **B** **C̶** **D** **Ⓔ** **F**

G-34

Connecting Math Concepts   Cumulative Test 2   63

Connecting Math Concepts   **Teacher's Guide   21**

Placement Testing

CMC Level A is appropriate for children who meet the placement criteria. The *Level A* Placement Test is used to measure children's abilities to follow oral directions.

A reproducible copy of the Placement Test Summary Sheet and the test for *Level A* appears on the following pages. The test consists of teacher's instructions (on the summary sheet) and child's sheet. The child's sheet can be reused since it does not get marked. The Summary Sheet, however, does get marked. Depending on the size of your class, you will need one to three Summary Sheets.

The test is administered to children individually, not to groups of children. Administration takes less than 1 minute per child.

Administering the Placement Test

Arrange to test children in a place that is reasonably quiet. The test may be administered by parents or volunteers.

The administrator is to:

- Fill out the child's name at the top of the summary sheet
- Present items 1–11 as specified
- Write + or – to indicate pass or fail
- Write the total number of errors
- Circle P (for 3 or fewer errors) or NP (for 4 or more errors)

Children must pass 8 of the 11 items in order to place in Level A. If the child does not pass the test, the child should not be placed in Level A. The child needs to receive more language instruction before beginning the program. (A possible program is *Language for Learning*.) After the children's language skills improve, they can be retested.

CMC Level A Placement Test Summary Sheet

Names	**Record + or − for each item**												
(Point to the first picture, of a man.)													
1. Is this a man or a woman? *Man.*													
2. Is this man wearing a coat? *No or head shake no.*													
3. Is this man wearing both shoes? *No or shake head no.*													
4. Is this man wearing a shirt? *Yes or head shake yes.*													
5. Touch his shirt. *Child touches man's shirt.*													
6. Touch his shoe. *Child touches man's shoe.*													
7. Show me where his other shoe would be. *Child touches man's other foot.*													
(Point to next picture, of ice-cream cone.)													
8. What is this? *Ice cream or ice-cream cone.*													
(Point to last picture, of glass.)													
9. What is this? *Glass or cup.*													
10. Do you drink from a cup? *Yes or head shake yes.*													
11. Do you drink water from an ice-cream cone? *No or head shake no.*													
Total Errors													
Passing Criterion: 3 or lower error Circle P or NP	P NP	P NP	P NP	P NP	P NP	P NP	P NP	P NP	P NP	P NP	P NP	P NP	P NP

(Point to first picture.)

1. Is this a man or a woman?
2. Is this man wearing a coat?
3. Is this man wearing both shoes?
4. Is this man wearing a shirt?
5. Touch his shirt.
6. Touch his shoe.
7. Show me where his other shoe should be.

(Point to next picture.)

8. What is this?

(Point to last picture.)

9. What is this?
10. Do you drink from a cup?
11. Do you drink water from an ice-cream cone?

Teaching Effectively

Level A is designed to be presented to small groups up to an entire class. You should generally be able to teach one lesson during a 30- to 40-minute period. Children's independent work requires 5 to 15 minutes. The independent work can be scheduled at another time during the day.

Organization

If possible, form three small homogeneous groups of 10 or fewer children. See Implementing the Program, **Grouping**, for more details. The program will run far more smoothly if you follow these steps in setting up the group.

1. **Seat the children in a semicircle in front of you.** (This is possible if you have a small group.) Sit so that you can observe every child in the group, as well as the other members of the class who are engaged in independent work.

2. **Place the lowest performers directly in front of you.** (Do this regardless of size of group.) Place these children in the first row if there is more than one row. Seat the highest performers on the ends of the group (or in the last row). You will naturally look most frequently at the children seated directly in front of you. You want to teach until each child is firm. If you are constantly looking at the low performers, you will be in a position to know when they are firm. When the lowest performers are firm, the rest of the group will be firm.

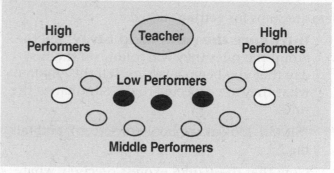

3. **Seat the children so cliques are broken.** Assign the seats. The children should sit in their assigned seat each day. This will allow you to separate disruptive buddies and allow you to learn which voices to listen to during the presentation.

4. **Test to see that all children can see the display.** Do this by holding your head next to the book and looking to see whether you can see the eyes of all the children. If you have to look almost sideways from the book to see a child's eyes, the child won't be able to see what is on the page.

If you are using the Board Displays, test the setup before class and check for visibility when arranging the seats. Keep in mind the children seated behind classmates will need to look around or above their fellow students.

Teaching

GETTING INTO THE LESSON

Here are steps for getting started.

1. **Introduce the rules right away.** Tell the group the rules they will follow on the first day that you begin a lesson. Tell the children what they are expected to do. Summarize the rules:

    "Sit tall, look at the book (or screen), and talk big."

    Note that these rules express precisely what the children are supposed to do. Reinforce them for following the rules.

2. **Get into the lesson *quickly*.** If the group is shy or tends to present behavior problems, begin by directing the children to stand or sit. "Stand up...sit down..." until all of them are responding without hesitation. This activity gets the children responding and establishes you as directing what they are to do. Then quickly present the first exercise. The same technique can be used if the children's attention lags during the presentation. It will break the pace and again establish you as directing what the children are to do.

3. **Present each exercise exactly as it appears in the presentation book.** If you change the vocabulary of a particular format, the children may have difficulty on future exercises. If you change a step in a format, the children may have trouble on a later exercise in which the step appears.

4. **Present each task until the children are firm.** This establishes what your criterion of performance is. Further information on **Teaching to Criterion** appears later in this section.

5. **Use clear signals.** All signals have the same purpose: to give the children a moment to think and to trigger a simultaneous response from the group. All signals have the same rationale: If you can get the group to respond simultaneously (with no child leading the others), you will get information about the performance of all the children, not just those who happen to answer first. The only alternative that gives you as much information about the children is to give **each** child an individual turn on each task. Testing each child takes a great deal of time and could promote behavior problems among those children who are waiting for their turns. Limit Individual Turns to a sampling of children. Use clear group signals to get clear feedback on the group.

    The specific signals used in the program are discussed later in the section on **Specific Signals.** Practice the signals until they are natural and you can do them without concentrating on them. The execution of a clear, easy-to-follow signal will result in efficient teaching of all the tasks.

6. **Pace tasks appropriately.** Pacing is one of the more difficult presentational skills to master. Pacing is the rate at which different parts of the exercises are presented. All portions of an exercise should not be presented at the same rate. The formats contain instructions to pause for different lengths of time (1 to 3 seconds). (For example, the pauses between numbers when children count from 1 to 10 should not be as long as the pauses between numbers children count when they are making a specific number of lines in the Workbook).

    Here are some general guidelines for pacing:

    - Speak quickly and with expression.

    - Stress words that are important by saying them loudly—not slowly.

    - Follow the instructions for pausing that appear in the Presentation Book.

    - Repeat a task if your pacing is poor. If you miss a line or say it too slowly, if you put an example on the board too slowly, if you accidently skip a display, if you rush a signal, repeat the part of the task that you fumbled. Tell the children, "Stop. Let's try that again."

Move from one task to the next quickly, pausing no longer than 3 or 4 seconds. As soon as the children are firm on one task, praise them. Then say, "New problem," and present the next task. From time to time you can take a brief "time out" with the children if necessary. More information appears later in **Pacing Your Presentation**.

7. **React to the children's responses.** Perhaps the most difficult skill to master is that of responding to the children's responses. The general rule is that if the children respond appropriately, you acknowledge it. For example, you ask, "What's 4 plus 2?" The children say, "6." And you say, "Yes, 6." If they have completed an entire task successfully, praise them. When they make mistakes, correct them immediately and effectively.

REINFORCING GOOD PERFORMANCE

1. **Reinforce when appropriate.** The group is performing well and deserves reinforcement when the following apply:
   - All children respond together on signal
   - All children give the correct response

2. **Reinforce children when they go through *all* of the steps in an exercise without making a mistake.** The exercise is designed to teach a skill, and the skill will be mastered only when responses are correct for every step in the exercise.

3. **Tell the children the goal they are working toward.** Say, "Here's a hard problem. Let's see if everybody can work the whole thing."

4. **Tell the children why you are praising them.** After the children have worked the whole problem correctly, say, "That was good. You worked the whole thing. And that problem was tough!"

5. **Do not spend a great deal of time reinforcing the children.** Take a moment to praise them, and then move quickly to the next problem. "This problem's even harder. Let's see if you can do it."

6. **Challenge the children.** "This next problem, I think it's too tough for you." A challenge often motivates the disinterested child to become an eager participant.

This challenge works best if most of the children will probably be able to work the problem correctly.

If you think fewer than half of the children will perform the task correctly, do not present the challenge and DO NOT present the task. Instead, firm the parts of the task children have trouble with. After children are firm on the parts, and you have confidence that most of the children will perform the task correctly, present the task and the challenge.

Using this type of challenge when children are not able to perform correctly is not reinforcing. The challenge simply confirms that the material is too hard for children.

7. **Use tangible reinforcers if the children do not respond well to verbal praise.** Use crackers, raisins, stars, points accumulated toward a small toy—something the children are willing to work for. If you have to use tangible rewards, always tell the children why they are receiving them. Say "Alex can count to 10," as you hand her a raisin.

8. **Reinforce children only when they perform according to an acceptable standard.** If you reward an individual child for poor performance, that child isn't learning. Furthermore, you will lose your credibility with the other children in the group.

Using the Teacher Presentation Script

The script for each lesson indicates precisely how to present each structured activity. The lesson is a script that shows you what to say, what you do. And what the children's responses are.

Follow the specified wording in the script. While wording variations from the specified script are not always dangerous, you will be assured of communicating clearly with the children if you follow the script exactly. The reason is that the wording is controlled, and the tasks are arranged so they provide succinct wording and focus clearly on important aspects of what the children are to do. Although you may feel uncomfortable "reading" from a script (and you may feel that the children will not pay attention), follow the scripts very closely; try to present them as if you're saying something important to the children. If you do, you'll find after a while that working from a script is not difficult and that children will indeed respond well to what you say.

SCRIPT CONVENTIONS

What you say appears in blue type:

> You say this.

What you do appears in parentheses:

> (You do this.)

The verbal responses of the children are in italics:

> *Children say this.*

What you say with children is in blue bold italics:

> **You and the children say this.**

The descriptions of children's actions appear in parentheses and are in italics:

> *(Children do this.)*

Three letters in a row indicate parts of words that are held for more than 1 second.

You hold the E sound in *nineteeen* for more than a second.

Children hold the short I sound in *siiix* for more than a second.

TEACHER DIRECTIONS

There are 5 main types of teacher directions in the *CMC* Teacher Presentation scripts: Signals; Repeats; Displays; Individual Turns; Observations.

Sample exercises 1 and 2 from Lesson 41 and sample exercise 1 from Lesson 118 appear on the following page. These exercises contain most of the variations of 4 of the 5 types of teacher directions: Signals; Repeats; Displays; Individual Turns. A discussion of these 4 types of teacher directions follows the samples.

A discussion of the fifth type of teacher directions, Checking and Observing, appears later in this section after the heading Teaching to Criterion in the part **Pacing Your Presentation.**

Lesson 41

EXERCISE 1: ROTE COUNTING—*Count to 49*

a. I'll get numbers going and say the next number.
Listen: nineteeen, 20.
New number: twenty-niiine, 30.
New number: thirty-niiine, 40.
Now you'll answer some questions.

(1a)
Signal
(2a)

- Everybody, when you count to 29, what number comes next?
(Signal.) *30.*
- When you count to 39, what number comes next? (Signal.) *40.*
(Repeat until firm.)

(2b)

b. Everybody, start with 40 and count to 49.
Get it going. *Fortyyy.* Count. (Tap 9.) *41, 42, 43, 44, 45, 46, 47, 48, 49.*
(Repeat step b until firm.)

Tap **(1b)**

c. Everybody, start with 37 and count to 45.
Get it going. *Thirty-sevennn.* Count. (Tap 8.) *38, 39, 40, 41, 42, 43, 44, 45.*
(Repeat step c until firm.)

===== **INDIVIDUAL TURNS** =====

(4) (Call on individual children to start with 37 and count to 45.)

EXERCISE 2: TEEN NUMBERS

(3a)

a. (Write on the board:)

[41:2A]

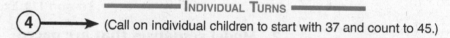

(1c)
Point and touch

Read these teen numbers.

- (Point to **14**.) Get ready. (Touch.) *14.*
- (Repeat for 15 and 16.)
(Repeat until firm.) **(2c)**

b. See if you can figure out a new teen number.

(3b) (Write to show:)

[41:2B]

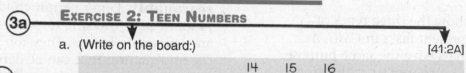

- (Point to **3**.) This is not a teen number.
What number is it? (Touch.) *3.*
(Write to show:)

[41:2C]

| 13 | 14 | 15 | 16 |

Now it's a teen number.

- (Point to **13**.) What teen number is it? (Touch.) *13.*
Yes, it's 13.

c. Everybody, read all these teen numbers.
- (Point to **13**.) Get ready. (Touch.) *13.*
- (Repeat for 14, 15, and 16.)
(Repeat step c until firm.)

Lesson 41, Exercises 1 and 2

Lesson 118

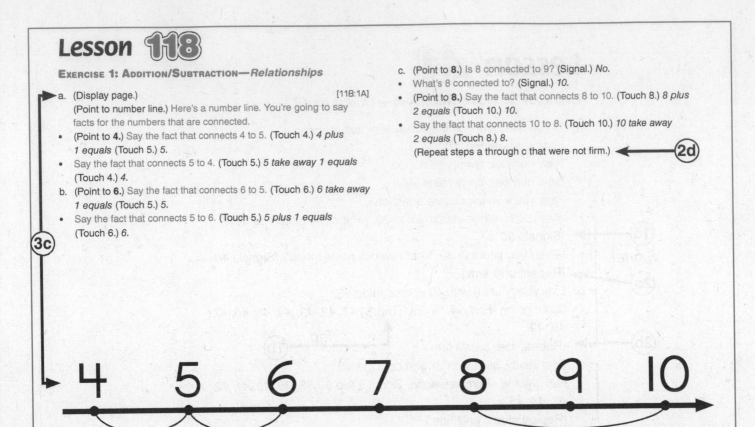

a. (Display page.)   [118:1A]
(Point to number line.) Here's a number line. You're going to say facts for the numbers that are connected.
- (Point to **4**.) Say the fact that connects 4 to 5. (Touch 4.) *4 plus 1 equals* (Touch 5.) *5.*
- Say the fact that connects 5 to 4. (Touch 5.) *5 take away 1 equals* (Touch 4.) *4.*
b. (Point to **6**.) Say the fact that connects 6 to 5. (Touch 6.) *6 take away 1 equals* (Touch 5.) *5.*
- Say the fact that connects 5 to 6. (Touch 5.) *5 plus 1 equals* (Touch 6.) *6.*

c. (Point to **8**.) Is 8 connected to 9? (Signal.) *No.*
- What's 8 connected to? (Signal.) *10.*
- (Point to **8**.) Say the fact that connects 8 to 10. (Touch 8.) *8 plus 2 equals* (Touch 10.) *10.*
- Say the fact that connects 10 to 8. (Touch 10.) *10 take away 2 equals* (Touch 8.) *8.*
(Repeat steps a through c that were not firm.)

Lesson 118, Exercise 1

Signals

Arrows 1a, 1b, and 1c show the three types of signals teachers use to present tasks in *CMC*. As indicated above, all signals have the same purpose: to trigger a simultaneous response from the group. All signals have the same rationale: If you can get the group to respond simultaneously (with no child leading the others), you will get information about the performance of all the children, not just those who happen to respond first.

The simplest way to signal children to respond together is to adopt a timing practice—just like the timing in a musical piece.

General Rules for Signals

Children will not be able to initiate responses together at the appropriate rate unless you follow these rules:

1. **Talk first.** Pause a standard length of time (possibly one second), then signal. Never signal when you talk. Don't change the timing for your signal. Make your signal predictable, so all children will know when to respond. Children are to respond on your signal—not after it or before it.

2. **Model responses that are paced reasonably.** Don't permit children to produce slow, droning responses. These are dangerous because they rob you of the information that can be derived from appropriate group responses. When children respond in a droning way, many of them are copying responses of others. If children are required to respond at a reasonable speaking rate, all children must initiate responses; therefore, it's relatively easy to determine which children are not responding and which are saying the wrong response.

Also, don't permit children to respond at a very fast rate or to "jump" your signal.

To correct mistakes, show the children exactly what you want them to do.

- I'm good at answering the right way.
- Watch: When you count to 39, what number comes next? 40.
- Let's see if you can answer that question at the right time.
- When you count to 39, what number comes next? (Signal.) *40.*

3. **Do not respond with the children (unless you are trying to work with them on a difficult response).** You present only what's in blue. You do not say the answers with the children unless the script specifies you to. You should not move your lips or give other spurious clues about what the answer is.

Think of signals this way: If you use them correctly, they provide you with much diagnostic information. A weak response suggests that you should repeat a task and provides information about which children may need more help. Signals are, therefore, important early in the program. After children have learned the routine, the children will be able to respond on cue with no signal. That will happen, however, only if you always give your signals at the end of a constant time interval after you complete what you say.

Basic Signal (Arrow 1a)

A basic signal follows a question, a direction, or the words, "Get ready." You can signal for arrow 1a by nodding, dropping your hand, clapping one time, snapping your fingers, or tapping your foot. After initially establishing the timing for signals, you can signal through voice inflection only. Signals specified by the direction (Signal.) may be visual or audible depending on where children's attention is focused. If children are focused on a display, or in their Workbook, a basic signal must be audible. If children are focused on you, the basic signal can be audible, but will work if it is only visual (nodding or dropping your hand).

Tap Signal (Arrow 1b)

A tap signal is shown by arrow 1b. Signals specified as (Tap.) must be audible signals, such as tapping your foot, snapping your fingers, or clicking a clicker. When children count together you'll lead them with audible serial taps. When children are looking at their worksheet, your signal needs to be audible so they can keep their eyes focused on the book, not on you. The timing of serial taps is extremely important. The timing of the taps for specific tasks should be consistent. If you model a task and then direct children to perform it, the timing of the serial taps should be identical to the rate at which you modeled it.

Point-and-Touch Signal (Arrow 1c)

The signal for arrow 1c is a point-and-touch signal. This signal is used in connection with a display. There are two teacher directions for every point-and-touch signal.

Pointing:

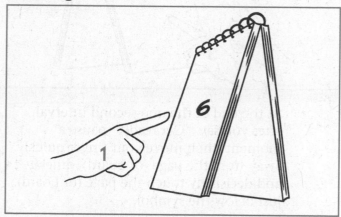

- Hold your finger about an inch from the page (or the board), just below the symbol.

- Be careful not to cover the symbol— all the children must be able to see it.

- Hold your finger in the pointing position for at least one second.

- Say, "Get ready." (In some formats, "Get ready" is replaced with a question such as, "What's this?")

- If necessary, look at the symbol to demonstrate to the children that they too should be looking at it.

Touching:

- At the end of the one-second interval, after you say, "Get ready," pause a moment, then move your finger quickly away from the page (or board), quickly and decisively touch the page (or board) just below the symbol.

- The instant your finger touches the display, look at the lowest-performing child in the group. See if the child responded the instant you touched the symbol.

Common pointing errors:
- Touching the symbol instead of pointing to it
- Pointing to another symbol or not clearly pointing to the target symbol
- Covering the symbol so children can't see it
- Pointing for less than one second
- Not saying "Get ready" or asking the specified question

Common touching errors:
- Touching the symbol before the one-second interval
- Touching the symbol while you're talking
- Touching the symbol indecisively or out of rhythm
- Pulling your finger too far from the page, making children uncertain which symbol you'll touch
- Covering the symbol while touching it
- Failing to look at the low performer the instant you touch the symbol

Repeats

There are two categories of repeats. One kind of repeat alerts you to the importance of the tasks that are taught, and implies a special correction if children make a mistake. The other kind of repeat indicates how to present additional examples. Arrows 2a, 2b, and 2d show where you may need to make a special correction. Arrow 2c shows the task and the examples that should be repeated.

A special correction is needed when correcting mistakes on tasks that teach a sequence or a relationship. This type of correction is marked with one of three notes:

- (Repeat until firm.) Illustrated by arrow 2a.
- (Repeat step ___ until firm.) Illustrated by arrow 2b.
- (Repeat steps that were not firm.) Illustrated by arrow 2d.

The repeat directions appear at the end of the tasks that are to be repeated. The tasks and the direction are bracketed on the left to clarify the tasks that need to be repeated. The tasks must be mastered before material that follows is presented. Repeat-until-firm and repeat-step-_-until-firm directions are used: after a sequence of tasks teaching a relationship that children may not understand if each of the tasks aren't responded to correctly and fluently; when children must produce a series of responses in a consistent sequence (as in counting).

For (Repeat until firm.) and (Repeat step _ until firm.) follow these steps:

1. Correct the mistake. (Tell the answer and repeat the task that was missed.)

2. Return to the beginning of the specified step and present the entire step.

In Exercise 1, step A, you present two different problems. For each problem, you ask, "When you count to __, what number comes next?"

If children are confused, if some don't answer either question, or if the answers to both questions are wrong, you repeat the bracketed part of step A after you have corrected.

When you hear a mistake, you say the correct answer and repeat the question or task. However, you make sure that children are firm in both the problems you present in step A. You can't be sure that children are firm unless you repeat the step.

Here's a summary of the steps you follow when repeating a part of the exercise until firm.

Correct the mistake:

(Tell the correct answer.) It's 30.

Repeat the task:

Listen again. When you count to 29, what number comes next?

Repeat the step:

Let's see who remembers what number comes next.

(Start at the beginning of the bracket in step A and present the entire task.)

"Repeating until firm" is based on the information you need about the children. When the children made the mistake, you told the answer. Did they remember the answer? Would they now be able to perform the procedure or sequence of responses correctly? The repeat-until-firm procedure provides you with answers to these questions. Children show you through their responses whether or not the correction worked, whether or not they are firm.

Lesson 41, Exercise 1, step A, shows a (Repeat until firm) direction with the arrow 2a. This repeat-until-firm direction firms a sequence of tasks teaching a relationship. The answers to the next-number questions are the numbers for counting by 10. If children don't answer the questions fluently and correctly, they may not connect the relationship between the numbers for counting by tens and numbers that are one more than quantities with a ones digit of 9.

A (Repeat step _ until firm.) direction occurs when the series of tasks to be repeated involve an entire step or steps. Arrow 2b directs you to repeat Lesson 41, Exercise 1, step B, until children start with 40 and count to 49 without making a mistake. This repeat direction firms a series of responses that must be produced in a consistent sequence.

A (Repeat _____ that were not firm) direction occurs when children are expected to apply procedures to a set of examples. Arrow 2d directs you to repeat the steps in Exercise 1 of Lesson 118 that required you to make a correction. Only the examples that children miss should be repeated when this direction appears. Do not repeat examples children responded to correctly.

Arrow 2c shows a task that should be repeated for the indicated examples. The arrow directs you to point to 15, say, "Get ready," and touch 15. Children should say, "15," after you touch it. These actions are to be repeated with 16. Arrow 2c shows a simple version of the repeat, indicating how to present additional examples.

Displays and Write to Shows

Arrow 3a shows the direction (Write on the board). Arrow 3b shows the direction (Write to show). There are various ways you can follow each direction. A write-on-the-board direction indicates that you can write the display on a white board or a chalk board. You can also project the Board Displays from the ConnectED on a screen or an interactive white board.

What you display after a write-to-show direction is a modification of the previous display. The write-to-show directions can be modified in the medium that you decide to show the original display. Usually, the changes in a write-to-show display are highlighted in the picture that appears after the direction.

Arrow 3a also points to the code that appears in brackets. The code that appears after each display corresponds to the frame for the display on the Board Displays. See pages 4 and 6 for information about the features of and how to use the Board Displays.

Arrow 3c shows the direction (Display page.) A visible version of display appears on the teacher presentation page. The display can also be projected using the corresponding code from the Board Displays.

Individual Turns

Individual Turns are specified in the exercises under the heading *Individual Turns*. Arrow 4 shows an Individual Turn. Individual Turns usually appear at the end of exercises.

There are several rules to follow when administering individual turns:

1. **Present Individual Turns only after the group is firm.** If you go to Individual Turns too soon, many of the children will not be able to give a firm response. If you wait until the children are firm on group responses, the chances are much better that each will be able to give a firm response on an Individual Turn.

2. **Give most of your Individual Turns to the lowest-performing children.** The lowest performers in the group are those children seated directly in front of you. By watching these children during the group practice of the task, you can tell when they are ready to perform individually. When the lowest performers (that are appropriately placed in the program) can perform the task without further need of correction, you can safely assume that the other children in the group will be able to perform the task. Unless your group is small, do not call on every child; doing so can cause restlessness.

3. **When a child makes a mistake on an Individual Turn, firm the group.** If one individual in the group makes a particular mistake, there is probably one other child in the group who will make the same mistake. The most efficient remedy, therefore, is to firm the entire group. Then retest the child who made the mistake.

If children are consistently weak on Individual Turns, you can assume that children are not mastering the material during your presentation of the tasks to the group. Look for these response problems:

- Group responses that are not in unison
- Children not attending
- Children repeating mistakes after corrections
- Unmotivated children

The presentation areas that may be causing these problems are:

- Weak signals
- Poor pacing
- Ineffective corrections
- Weak reinforcement

Teaching to Criterion, which appears later in this section, contains information about how to fix these presentation areas.

4. **Don't skip Individual Turns.** Always include the Individual Turns for tasks in which they are specified. Individual Turns are not specified in all exercises. If you are in doubt about the performance of any children on a task, present quick individual turns.

Teaching to Criterion

At the conclusion of any exercise, every child should be able to perform each task independently, without any need for corrections. Children are "at criterion" or "firm" on a task only when they can perform immediately with the correct response. Your goal is to teach so that every child is at criterion.

It is easier to bring the children to criterion on the first introduction of a format than it will be at a later time in the program because they haven't performed the task incorrectly many times or heard others performing it incorrectly.

The initial formats in each track include a demonstration by the teacher of the response that the children are to make. This provision allows the children to hear the correct response the first time that the task is presented.

Let the children know what your criterion is. Stay on a task until you can honestly say to them, "Terrific. Everybody can work the problem correctly." The stricter your criterion, the fewer children you will have requiring additional practice after taking the Mastery Tests.

CORRECTIONS

The major difference between the average *CMC Level A* teacher who teaches a majority of her children and the teacher who has successfully taught *all* her children is the ability to correct.

Information on general corrections appears on the following pages. Study these procedures and practice them until you can execute them immediately, without hesitation. These corrections must become automatic. Failure to get each child to pay attention or allowing part of the group not to respond will result in some of the children not learning.

The **Tracks** section contains information about how to correct mistakes in specific exercises; however, the basic correction procedure can be used to correct any mistake. The basic correction may not be more efficient or effective than specific corrections described in the **Tracks** section, but it will work.

A general rule that applies to all corrections is to respond immediately when a mistake occurs.

Unacceptable behavior that calls for correction includes: nonattending, nonresponding, signal violations, and response errors.

1. **Nonattending.** This behavior occurs when a child is not looking where he should be looking during a task. For example, if a child is not attending to the numeral to which you are pointing, correct by looking at the non-attender and saying:

   Watch where I'm pointing.

   Let's try it again.

   Return to the beginning of the task.

   Reinforce the children who are paying attention. Let them know you are watching all of them all the time. Always return to the beginning of the task to enforce your rule that everyone has to pay attention at all times.

2. **Nonresponding.** This behavior occurs when a child fails to answer when the teacher signals a response. It is dangerous to overlook nonresponding. The children may learn to just listen for the answer the first time a task is presented and then respond correctly with the group when the task is repeated. Children who don't respond the first time a task is presented will learn dependence on other children and get the idea that they need not answer along with the rest of the group. If a child is not responding, correct him by saying:

   I like the way some of you said it the right way. But I need everyone to do it. Let's do that again.

   Return to the beginning of the task.

   Failure to return to the beginning of the task will teach the children that you really do not mean that you "need everyone to do it." It is very important to enforce this rule from the first day of instruction so that the children learn you are expecting everyone to perform on every task.

3. **Signal Violations.** A signal violation occurs when the child responds either before or too long after the signal or during the portion of the task in which you are demonstrating. For example, a child might begin to identify a numeral after you have touched it and after the other children respond. The lower-performing children are most likely to violate the signal because they will tend to wait for the higher-performing children in a group to respond first.

   If children respond either early or late, you will not get information from every child. Remember that the purpose of a signal is to trigger a simultaneous group response. If you fail to enforce the signal, you will not get information about which children are firm and which children are weak.

   Only if you consistently return to the beginning of the task after each signal-violation correction will the children learn to attend to your signal. Once they learn that you will repeat the task until they are all responding on signal, they will attend much more closely to the signal.

   If you find that you are spending a lot of time correcting nonattending, nonresponding, and signal violations, your pacing is probably too slow, or your pacing of the signal inconsistent. The object of a signal is not to keep the children sitting on the edges of their seats, never knowing when they will have to respond next. The pacing of the signal should be perfectly predictable so all children know when they are expected to respond.

4. **Response Errors.** A response error is any response inconsistent with the one called for in the task. If you are teaching symbol identification, and the children say *4* when you touch 2, this is a response error. They may have followed your signal and identified the numeral just when you touched it, but their response is inconsistent with the symbol you are pointing to.

   Response errors are specific to the individual task. The correction for each response error, therefore, must be specific. The children's mistakes can be anticipated. Occasionally corrections will appear on the page with the appropriate format in the Presentation Book.

The most common mistakes that children make are identified in the **Tracks** section, and the appropriate corrections are supplied. See Tracks, **Counting**, for an example of an embedded To-correct procedure (in step E of the Next-Number exercise from Lesson 11). It is very important to practice these corrections. You must be able to present the correction without hesitation when the mistake occurs. By practicing the corrections, you will be well prepared for the common mistakes that the children will make. It is important not to present the indicated correction if children do not make a mistake. It is also important to use that correction for any of the similar tasks, even though the correction is specified for one task.

When correcting a response error, always return to the beginning of the task (or to the lettered step specified in the To-correct section of the task) after you have corrected a mistake. This is very important because the children must learn that the various steps they take in learning a skill are not isolated—they fit together in a sequence. The children will not become familiar with the sequence unless they respond correctly to the steps presented in the sequence.

The first formats in almost every track are written so that the teacher first demonstrates, or models, the response the children are to make. Sometimes the next step is a *teacher lead*, in which the teacher responds with the children. Leading is a very powerful technique. It gives the children the benefit of responding with you until they are confident in the pace and responses they are to make. Many tasks require a number of teacher leads before the children are producing firm responses. The lead is a correction. Don't be afraid to continue leading until the children are producing the response with you. Only after they can produce the response with you will they be able to produce it on their own.

Don't be afraid to model a task (such as saying a sequence of new numbers) two or three times before directing children to say it. However, a general rule is that if children are not able to perform on a task (such as counting) after about seven trials, don't continue to repeat the task. We don't want to grind children. After the seven trials, model the sequence once more, then tell the children, "We'll work more on that later."

Repeat the instruction for the task children were unable to perform after presenting at least one or two more exercises. Returning to a task after intervening instruction is called a delayed test.

If children make several mistakes on an exercise, provide a delayed test on the entire exercise after presenting a few more exercises.

The correction scenario for a very difficult task is as follows:

- Model: The teacher immediately models the correct response.

- Lead: (Not all corrections need this step.) Teacher performs the task with the students.

- Test: Repeat the task to test performance.

- Retest: Go to the beginning of the step or exercise and retest.

- (Delayed test): Go back to the step or exercise after intervening instruction and retest.

Pacing Your Presentation— Checking and Observing

You should pace your verbal presentation at a normal speaking rate—as if you were telling someone something important.

CMC Level A uses 3 teacher directions to pace your presentation for activities in which students write, work problems in their heads, or touch and find parts of their Workbooks. The first is a ✔. The second is a note to (Observe children and give feedback.) The third is a note to (Pause).

Check and Observe indicate that you will interact with children. Some interactions will serve to correct mistakes. Others reinforce desired behaviors. In other words, ✔ and (Observe children and give feedback.) are signals for managing children and giving feedback that helps the children perform better.

A ✔ is a note to check what the students are doing. It requires only a second or two. If you are positioned close to several lower-performing students, check whether they are responding appropriately. If they are, proceed with the presentation. If they aren't, correct the group.

The (Observe children and give feedback.) direction implies a more elaborate response. You sample more children and you give feedback, not only to individual children but to the group. Here are the basic rules for what to do and what not to do when you observe and give feedback:

1. Move from the front of the room to a place where you can quickly sample the performance of low, middle, and high performers.

2. As soon as students start to work, start observing. As you observe, make comments to the whole class. Focus these comments on children who are following directions, working quickly, and working accurately. "Wow, a couple of students are almost finished. I haven't seen one mistake so far."

3. If you observe mistakes, do **not** provide a great deal of individual help. Point out any mistake, but do not work the problems for the students. For instance, if a child gets one of the problems wrong, point to it and say, "You made a mistake." If children don't line up their numerals correctly, say, "You'd better erase that and try again. Your numbers are not lined up." If students are not following instructions that you give, tell them, "You didn't follow my directions. You have to listen carefully. Just make the lines, then stop."

4. If you observe a serious problem that is not unique to the lowest performers, tell the class, "Stop. We seem to have a problem." Point out the mistake. Repeat the part of the exercise that gives them information about what they are to do. *Note:* Do not provide new teaching or new problems. Simply repeat the part of the exercise that gives children the information they need and reassign the work. "Let's see who can get it this time."

5. Do not wait for the slowest child to complete the problems before presenting the workcheck during which children correct their work and fix any mistakes. Allow children a **reasonable** amount of time. You can usually use the middle performers as a gauge for what is reasonable and then observe and give feedback for a few more seconds before proceeding.

On Mastery Tests you will sometimes see the direction (Observe children.). For this direction, you observe children as they perform written work. You do not give them specific feedback. However, we recommend that you use the time to mark children's papers (make a C next to problems that are worked correctly and an X next to problems that are not worked correctly) if you are not presenting Individually Administered portions of the test.

(Pause.) is a direction for you to give children a second or so to think. The amount of time you pause should be adjusted based on children's responses. If children all respond correctly, you may be able to shorten the time you pause. If children are not responding correctly or in unison, make the time you pause longer.

Practice Tasks

All of the teacher behaviors discussed so far are essential to teaching every task in the program. The Addition and Subtraction exercises that follow should be used during preservice training to practice these critical behaviors.

Follow the task script exactly as it appears, pacing your presentation appropriately. Execute the signals clearly and react to the responses. Make corrections effectively and reinforce good performance when the group is at criterion.

The adults will play the role of the children, and some adults will make the mistakes (response errors) that are suggested. At any point in your presentation, some adults may make unscripted mistakes which require you to correct for nonattending, nonresponding, or signal violations. Repeat the task until all the adults can go through the entire exercise without errors.

The later objectives of the program determine what is taught earlier so that students will have the skill and knowledge necessary to meet all the objectives.

By Lesson 95, children work addition and subtraction problems that have two-digit numbers. These problems show what children have learned earlier to prepare them for the work on these problems. Basically, every component skill in the operations presented is taught earlier—first in isolation and then in more complex operations.

We will briefly look at the various objectives that are implied by the operations children learn for addition and (in the section that follows this) subtraction.

Practice Task—Addition

Below is an addition exercise like those presented around Lesson 95.

Read this exercise out loud at least three times before proceeding. Say the teacher directions and say the children's oral responses.

After you are familiar with the exercise, continue reading this section.

a. Touch the tiger on your worksheet. ✔
    (Teacher reference:)

$$37$$
$$+\ 24$$

b. Touch and read the first problem. (Signal.)
    37 plus 24.

c. You'll make counters for one of the numbers.

• Which number? (Signal.) *24.*

d. How many Ts will you make for 24?
    (Signal.) *2.*

• How many lines will you make for 24?
    (Signal.) *4.*

• Make the Ts and lines for 24.
    (Observe children and give feedback.)
    (Teacher reference:)

$$37$$
$$+\ 24\ \text{TT||||}$$

e. Now you'll count for both groups.

• Touch the number you'll get going. ✔

• Get it going. *Thirty-sevennn.* Count for the Ts. (Tap 2.) *47, fifty-sevennn.* Count for the lines. (Tap 4.) *58, 59, 60, 61.*
    (Repeat until firm.)

f. What's the answer to the problem?
    (Signal.) *61.*

g. Write 61. ✔
    (Teacher reference:)

$$37$$
$$+\ 24\ \text{TT||||}$$
$$61$$

h. Touch and read the equation. (Tap 5.)
    37 plus 24 equals 61.

As noted on the previous page, examining the steps of the exercise reveals the skills that are taught earlier in Level A.

Step A: Touch the tiger on your worksheet. ✔

Step A instructs children to touch the tiger on their worksheet. Early in the program children are taught to locate parts on their worksheet by locating pictures of familiar objects. Objects that may not be familiar to children, like tigers, are taught to children before they are required to locate them on worksheets.

Step B: Touch and read the first problem.
(Signal.) *37 plus 24.*

The children's response indicates that they are able to decode two-digit numbers and the sign for plus (+). Their response also implies that they have learned to read symbols written in rows or columns in the appropriate order and are already able to identify single-digit numbers from earlier in the program.

Step C: You'll make counters for one of the numbers.
• Which number? (Signal.) *24.*

The children's response to this step shows that they have been taught a procedure or algorithm for solving addition problems. They have learned that they make counters for the number that is added. In this case they make counters for 24 and do not make counters for 37.

Step D: How many Ts will you make for 24?
(Signal.) *2.*
• How many lines will you make for 24?
(Signal.) *4.*
• Make the Ts and lines for 24.
(Observe that children write **T T** I I I I
next to 24 and give feedback.)

Children who can perform the tasks in step D know how to decompose a two-digit number like 24 into tens and ones. They have learned that the tens digit tells them to make 2 Ts (counters

for tens), and the ones digit tells them to make 4 lines (counters for ones). In other words, children not only understand place value, but are able to represent place value with counters for tens and ones. Children's responses indicate that they have also learned to make counters for numbers: counters for one-digit numbers; counters for decade numbers; and counters for other two-digit numbers.

Step E: Now you'll count for both groups.
• Touch the number you'll get going.
(Check that children touch 37.)
• Get it going. *Thirty-sevennn.*
• Count for the Ts. (Tap twice; children touch Ts as they count:) *47, fifty-sevennn.*
• Count for the lines. (Tap 4 times; children touch lines as they count:) *58, 59, 60, 61.*

Error Suggestions:

Nonattending; nonresponding (not touching 37); signal violation; response error (such as 30 or 24 instead of 37, 38 instead of 47, 67 instead of 58)

This step involves a complicated count-on strategy. Children know that they start with 37 (the value of the first number). But they don't count until they touch the first T. The fact that children are able to start with 37 and add 10 for each T implies that they are practiced in adding tens to various two-digit numbers.

Note that children have also been taught the discrimination between adding 10 for each T and adding 1 for each line. Note also that adding 1 is linked to the last number for adding Ts: 57 (which is followed by 58).

Specifically, children earlier learn the component counting skill of starting with a range of two-digit numbers and adding either tens or ones. This counting skill implies mastery of counting by ones, by tens, and counting on by adding ones or adding tens.

Step E: (Repeat until firm.)

Teach step E to criterion. The bracket guides you to repeat the task from the beginning of step E.

Step F: What's the answer to the problem?
(Signal.) *61.*

The children's response implies that they have learned that the last number said after a counting sequence is the total for what is counted and is the answer.

Step G: Write 61. ✔

$$
\begin{array}{r}
37 \\
+\ 24 \quad \text{TT}||||\ \\
\hline
61
\end{array}
$$

The children's response implies that they are practiced at writing two-digit numbers. Children have also learned to write one-digit and two-digit numbers. Children are practiced at aligning tens digits below tens digits and ones digits below ones in column problems. The ✔ symbol directs the teacher to look up from the presentation book to check that children are writing.

Step H: Touch and read the equation. (Tap 5.)
37 plus 24 equals 61.

Children's responses indicate that they can read + as plus and _____ as equals, and they are familiar with the convention that completed problems in which both sides are equal are called **equations**. The response also implies that children know how to discriminate between the answer to a problem (steps F and G) and an equation that contains the answer (step H). The teacher taps 5 times, once for each component in the equation: 37, plus, 24, equals, 61.

Practice Task—Subtraction

Below is a model of the routine children follow to work two-digit subtraction problems. The procedure children use is different from the addition procedure. For subtraction, children make counters for the first number in the problem, not the number subtracted. They show the value that is subtracted by crossing out counters. The counters that are not crossed out represent the answer to the problem.

Read this whole exercise out loud at least three times. Say the lines for the teacher and the children. **Before reading further in the guide, make sure you have a good understanding of the order of the steps, what you say, and the children's responses.**

a. Touch the problem 45 take away 32. ✔
(Teacher reference:)

$$
\begin{array}{r}
45 \\
-\ 32
\end{array}
$$

- Touch and read the problem. Get ready.
(Signal.) *45 take away 32.*

b. You'll make counters for one of the numbers.
- Which number? (Signal.) *45.*
- How many Ts will you make for 45?
(Signal.) *4.*
- How many lines will you make for 45?
(Signal.) *5.*
- Make the Ts and lines for 45.
(Observe children and give feedback.)
(Teacher reference:)

$$
\begin{array}{r}
45 \quad \text{TTTT}|||||\ \\
-\ 32
\end{array}
$$

c. Touch the number you take away in this problem. ✔
- What number do you take away?
(Signal.) *32.*
- How many Ts will you take away?
(Signal.) *3.*

- How many lines will you take away? (Signal.) *2.*
- Cross out the Ts and lines you take away. (Observe children and give feedback.) (Teacher reference:)

$$45 \; \text{T̶T̶T̶T̶ I I I H̶I̶}$$
$$- \; 32$$

d. Now you'll count for the Ts and lines that are not crossed out.
- Count for the Ts. (Tap 1.) *Tennn.* Count for the lines. (Tap 3.) *11, 12, 13.* (Repeat until firm.)

e. What's the answer to the problem? (Signal.) *13.*
- Write 13. ✔
- Touch and read the equation. (Tap 5.) *45 take away 32 equals 13.*

Some skills students use to perform the subtraction operation are the same as those for the addition operation. We will focus on the details that are different.

Step A: Touch the problem ___. ✔
- Touch and read the problem. Get ready. (Signal.)

The skills implied by the tasks children perform in step A are covered by the addition operation: finding a problem in their workbook, identifying symbols, and reading symbols in order.

Note that children have been taught to read the – sign as "take away," not "minus."

Step B: You'll make counters for one of the numbers.
- Which number? (Signal.) *45.*
- How many Ts will you make for 45? (Signal.) *4.*
- How many lines will you make for 45? (Signal.) *5.*
- Make the Ts and lines for 45. (Observe children and give feedback.) (Teacher reference:)

$$45 \; \text{T T T T I I I I I}$$
$$- \; 32$$

The critical difference between step B of subtraction and steps C and D of addition is that for subtraction children make counters *for the first number, not the second.*

This step implies that earlier practice was provided to assure that children learned to work take-away problems involving one-digit numbers in isolation. After children mastered plus problems in isolation, and take-away problems in isolation, they received a great deal of practice to become facile in distinguishing between when to make counters for the first number (take-away problems) and when to make counters for the second number (plus problems).

The observe-children-and-give-feedback note directs the teacher to check that children are 1) writing, 2) writing counters next to the correct number (45), and 3) writing the correct number of Ts (4) and lines (5). Children who are not writing the counters correctly need to receive immediate feedback.

Step C: Touch the number you take away in this problem. ✔
- What number do you take away? (Signal.) *32.*
- How many Ts will you take away? (Signal.) *3.*
- How many lines will you take away? (Signal.) *2.*
- Cross out the Ts and lines you take away.

    (Observe children and give feedback.)
    (Teacher reference:)

$$45 \; \text{T̶T̶T̶T̶ IIIH}$$
$$- \; 32$$

Step C reveals other important differences between the operations for addition and subtraction. For subtraction, children don't add counters but cross them out. This discrimination also implies that children have worked with groups of problems, some of which add and others that subtract.

The first two tasks in step C, "Touch the number you take away in this problem" and "What number do you take away?" are straightforward for children who have mastered reading numbers and the symbols – (take away) and _____ (equals).

You will quickly check (✔) children are touching 32 in the first task. Their responses to the second task will tell you if they were touching the correct number.

The next tasks in step C, "How many Ts will you take away?" and "How many lines will you take away?" are simple extensions of place-value skills.

The final task of step C, "Cross out the Ts and lines you take away," requires children to be familiar with the conventions for crossing out. You will observe and give feedback for this task. The first take-away tasks involve only lines. Children learn the convention that you make lines left to right. For crossing out, children start

to the right of lines and cross them out right to left. This convention is extended when two-digit numbers are introduced and children cross out Ts and lines. Children start to the right of the last T and cross out the correct number of Ts going from right to left. Then children start to the right of the last line and cross out the correct number of lines going right to left:

$$45 \; \text{T̶T̶T̶ IIIH}$$
$$- \; 32$$

Step D: Now you'll count for the Ts and lines that are not crossed out.
- Count for the Ts. (Tap 1.) *Tennn.* Count for the lines. (Tap 3.) *11, 12, 13.* (Repeat until firm.)

The counting is simple but different from what children do with addition. For addition they start with the first number in the problem and count on as they add for the Ts and then add for lines. For subtraction, children do not count on. They simply count for the Ts and lines that are not crossed out: *Tennn, 11, 12, 13.*

Teach step D to criterion by repeating until firm.

Step E: What's the answer to the problem? (Signal.) *13.*
- Write 13. ✔
- Touch and read the equation. (Tap 5.) *45 take away 32 equals 13.*

Like addition, the last number children say when they count is the answer to the problem.

Also, like addition, children must be facile in reading and writing two-digit numbers and completing horizontal as well as vertical problems and equations.

PREREQUISITE SKILLS FOR ADDITION AND SUBTRACTION

For children to become more fluent in performing the discriminations and sequence of tasks for both addition and subtraction problems, they need considerable practice in the prerequisite skills for these operations. These prerequisite skills include a variety of counting skills and symbol skills.

Counting Skills

- Counting by ones
- Counting by tens
- Counting events (making a line, making a T, or crossing out a line or a T)
- Counting on by ones
- Counting on by tens
- Counting for Ts and lines
- Counting on for Ts and lines

Symbol Identification and Symbol Writing

- Reading all single-digit numbers
- Reading signs (+, =, –, _____)
- Reading two-digit numbers
- Reading problems in correct order (left to right for problems written in rows, top to bottom and left to right for column problems)
- Writing all signs and single-digit numbers
- Writing two-digit numbers
- Writing problems and equations
- Writing Ts and lines

As these prerequisite skills are mastered, they are gradually and systematically integrated to teach the problem-solving operations. Children work the first problems for each operation on the board as a group. The teacher guides children through the operational sequence with a series of questions and tasks. After children have become familiar with an operational sequence, children work problems involving the operation (like the addition and subtraction problems shown above) in their workbook. For the first workbook problems, the teacher prompts the operational sequence with the series of questions and tasks. Gradually, children receive less and less teacher direction to perform problems involving the operation. Before the end of the program, children are sufficiently familiar with the discriminations and the sequence of tasks that they are able to perform addition and subtraction problems without teacher direction.

Tracks

Note: Before reading further in the guide, make sure you have a good understanding of the order of the steps, what you say, and the children's responses in **Practice Task—Addition** and **Practice Task—Subtraction** in the preceding section of this guide.

This section provides more detail about the skills and knowledge children learn in *CMC Level A*. The skills are divided into tracks. Each track focuses on specific skills. **The Scope and Sequence Chart**, pages 8–9, shows the tracks of *CMC Level A*. The bars show the lesson on which each track begins and the lesson range for the track.

Each lesson is composed of exercises from 6 to 9 tracks that present various (sometimes unrelated) skills. In other words, children do not devote a lesson to a single topic or skill. Rather they spend a few minutes working on each exercise. As component skills are learned, they are merged into operations. For example, after children learn to recite the counting numbers, they learn to count objects and, later, events, like the number of times the teacher claps. The operations of counting objects and events are merged when children make lines and count them: The events are making the lines; the objects are the lines the children make.

Most of the tracks address the skills that children need to perform the operations of addition and subtraction. These skills are extended to word problems, which entail translating terms like "8 birds" into **8** and "5 more" into **plus 5**.

In addition to the tracks that address the two primary operations, *CMC Level A* has tracks that address standards not directly related to the main operations that are taught. These tracks include identifying two-dimensional shapes and three-dimensional objects, ordering numbers in a specified pattern, and grouping different items by count.

See the last section of this guide, *CMC Level A* and Common Core State Standards, for a discussion about how *CMC Level A* meets all the standards specified for K by the Common Core State Standards for Mathematics.

Counting—Lessons 1–120

Counting is the foundation skill for every math operation children are to learn. The Addition and Subtraction Practice Tasks in the preceding section illustrated some of the counting skills that children learn in *CMC Level A*.

Here are the types of counting skills that children learn in *CMC Level A*.

- Reciting counting numbers (rote counting)
- Counting objects (including coins)
- Counting events
- Count on (next number)
- Counting from–to (from a number to a number)
- Counting by tens
- Counting backward
- Ordinal counting
- Combining counting skills (such as counting and making lines at the same time)
- Extensions that require application of counting skills (such as identifying the next number in a counting series which leads to learning plus-1 facts)

A general teaching note for counting activities is that the goal should be to make activities reinforcing, like a game—like the lyrics of a song—not a chore.

Part of this strategy means that you don't simply provide enough practice so children are able to perform the counting activities with great concentration but enough practice so children recite the counting numbers and perform the counting operations with fluency and confidence.

The most effective way to achieve this goal is to model that counting is an enjoyable activity. Smile, keep a pace that children can follow, and don't be afraid to model a sequence of new numbers two or three times before directing children to say it.

Another general rule is that if children are not able to perform on a counting task after about seven trials, don't continue to repeat the task. After the seven trials, model the sequence once more. Then tell the children, "We'll work more on that hard counting later."

Repeat the instruction for the counting task children were unable to perform after presenting another exercise or two.

RECITING COUNTING NUMBERS (ROTE COUNTING)

The table below shows the lesson number on which children first learn to count to different numbers. During the lessons that do not introduce a new number, children review counting to familiar numbers.

As the table shows, children learn to count through 199.

Note that the usual span of numbers children count in a task is 12 or fewer. For example, children don't count from 1 to 56. They may count from 46 to 56.

Rote Counting to Familiar Numbers

By Lesson	1	2	7	9	10	14	16	17	22	32	33	34	36	37	39	40	41
Children count to	4	5	6	7	10	12	13	15	19	25	29	30	35	39	40	45	49

By Lesson	45	46	47	50	52	53	54	55	56	57	74	75	76	78	79	82
Children count to	50	54	59	61	69	75	80	90	99	100	105	110	115	160	185	199

Here's the first Rote Counting exercise in Lesson 1 of the program, in which children count to 4:

EXERCISE 1: ROTE COUNTING—*Count to 4*

> Note: (Do not display the page until step g.)

a. Listen to me count: 1, 2, 3. I ended up with 3.
• Once more: 1, 2, 3. What number did I end up with? (Signal.) *3.* Yes, 3.
(Repeat step a until firm—children's response is correct and on cue.)

b. Listen: 1, 2, 3, 4, 5, 6. What number did I end up with? (Signal.) *6.* Yes, 6.

c. Listen: 1, 2, 3, 4, 5. What number did I end up with? (Signal.) *5.*

d. Listen: 1, 2, 3, 4. What number did I end up with? (Signal.) *4.*
• Listen again: 1, 2, 3, 4. What number? (Signal.) *4.*

e. Let's all count and end up with 4. Every time I tap, you count. Get ready. (Tap 4 as you say with children): *1, 2, 3, 4.*
(Repeat step e until firm.)

f. Your turn: Count and end up with 4. Get ready. (Tap 4.) *1, 2, 3, 4.* (Repeat step f until firm.)
• Do it once more, and I'll show you a picture of 4. Count and end up with 4. Get ready. (Tap 4.) *1, 2, 3, 4.*

g. (Display page.)   [1:1A]
Here's a picture of 4. The picture shows 4 tigers.
What are these 4 things? (Signal.) *Tigers.*
Yes, these 4 things are tigers.
• Listen: I know how many tigers there are. 4. I'll count them. (Touch and count:) 1, 2, 3, 4.
• Everybody, how many tigers? (Signal.) *4.*
(Repeat step g until firm.)

This exercise introduces children to three aspects of counting:

- Reciting the numbers in a fixed order
- Identifying the number the teacher counts to
- Learning the relationship between the counting numbers and the number of objects shown, pairing each number name with one object

Teaching Note: In steps A through D, you model counting to different numbers: 3, 6, 5, 4.

After you count to a number, you ask children, "What number did I end up with?" These tasks acquaint children with the idea that the last number you say is the number you end up with. Later in the exercise, children will see that the last number also describes how many objects are in a group.

Make sure you always count at the same pace. The pace that you set is the pace you expect the children to follow. If you model counting too fast, some children will not be able to keep up with the pace. Try to count at the same pace you would walk slowly, not at your normal walking rate.

When you model counting, you may want to tap in time with each number you say. The rate you tap during your model should be identical to the rate you tap when signaling children to respond.

Remember, children are to respond at the same rate that you model the counting. If the pace you set is too fast for some children, it will take much longer for them to become fluent than if you present a pace that is slow enough for them.

This exercise indicates that you repeat step E until firm. In step E you count with the children. Do not repeat the step more than about seven times. If children can count to 4 with you after fewer than seven trials, present step F. If children are not firm, skip step F and move to step G. In this step, you count 4 tigers. Children answer "How many tigers?" with the last number you counted, 4.

Children receive more practice counting in Exercises 5 and 7. Exercise 5 is parallel to Exercise 1 except that the 4 objects in the group are skunks.

In Exercise 7, children review some of the tasks presented in Exercises 1 and 5. Then they are introduced to saying the next number.

Here's the exercise:

EXERCISE 7: NEXT NUMBER

a. Listen to me count: 1, 2, 3, 4, 5, 6. What number did I end up with? (Signal.) *6.*
Yes, I counted to 6.

- Listen: 1, 2. What number did I end up with? (Signal.) *2.*
Yes, I counted to 2.

- Listen: 1, 2, 3, 4. What number did I end up with? (Signal.) *4.*
Yes, I counted to 4.

b. Let's all count and end up with 4. (Tap 4 as you say with children:)
1, 2, 3, 4.
(Repeat step b until firm.)

c. I'm going to count and say the next number. Listen: 1.
What's the next number? 2.
What's the next number? 3.
What's the next number? 4.

d. Do it with me. 1. What's the next number? (Signal.) *2.*
- Your turn: 1. What's the next number? (Signal.) *2.*
(Repeat step d until firm.)

━━━━━━ **INDIVIDUAL TURNS** ━━━━━━
(Call on individual children to perform the following task.)

- Listen: 1. What's the next number? (Call on a child.) *2.*

Lesson 1, Exercise 7

In steps A and B of this exercise children again identify the number you end up with and again count to 4 with the teacher.

In the next part of the exercise, step C, you model identifying the next number.

After you identify numbers through 4, in step D you say "One" and ask children "What's the next number?"

After repeating this task with children and directing children to do it without your lead, you call on children to perform Individual Turns. See the section on **Individual Turns** in the previous section, Teaching Effectively, for critical information about Individual Turns.

COUNTING OBJECTS

Counting objects assumes that each number tells about a member of a group. The number for the last object counted is, therefore, the number for all members of the group. In Lesson 6, children learn an important feature of counting objects. It doesn't matter which object in a group you count first. If you count each object, the final number for the group will always be the same. (Note that this relationship does not hold for counting events. Events are counted in their order of occurrence.)

Here's the exercise for Lesson 6:

a   b

EXERCISE 3: OBJECT COUNTING

a. (Display page.)   [6:3A]
   You're going to count things in this picture.
   These are things you'll count. What are these things? (Signal.)
   Birds.
   Yes, birds.
b. (Point to **a.**) Let's start here and count the birds. Get ready. (Touch each bird as you count with children:) *1, 2, 3, 4.*
• How many birds? (Signal.) *4.*
c. (Point to **b.**) Let's start here and count the birds. Get ready. (Touch each bird as you count with children:) *1, 2, 3, 4.*
• How many birds? (Signal.) *4.*
   Yes, 4.
d. Yes, we can count from this side (point to **b**) or the other side (point to **a**). We still end up with 4 birds.
• All by yourself: (Point to **a.**) Start here and count the birds. Get ready. (Touch each bird as children count.) *1, 2, 3, 4.*
• How many birds? (Signal.) *4.*
   Yes, 4.

e. Here are some more birds.
• (Point to **c.**) Let's start here and count the birds. Get ready. (Touch each bird and count with children:) *1, 2, 3, 4, 5.*
• How many birds? (Signal.) *5.*
f. (Point to **d.**) Let's start here and count the birds. Get ready. (Touch each bird and count with children:) *1, 2, 3, 4, 5.*
• How many birds? (Signal.) *5.*
• (Keep pointing to **c.**) Yes, we can count from this side or the other side. We still end up with 5 birds.
g. (Keep pointing to **c.**) All by yourself: Count the birds. Get ready. (Touch each bird as children count.) *1, 2, 3, 4, 5.*
• How many birds? (Signal.) *5.*
• (Keep pointing to **d.**) Yes, you can count from this side or the other side. You still end up with 5 birds.
h. (Keep pointing to **d.**) Let's see if you end up with 5 when we count from this side. Get ready. (Touch each bird as children count.) *1, 2, 3, 4, 5.*
• How many birds? (Signal.) *5.*

c   d

Lesson 6, Exercise 3

COUNTING EVENTS

When children count as you touch objects, they are actually counting events. Counting the number of times the teacher claps is a variation of counting events. This variation does not count anything permanent. The clap occurs and it is counted but no longer exists.

Children count claps but also count variations, such as pennies dropping in a cup. These variations are different because each event is counted when it occurs. But the evidence that the teacher dropped 5 pennies, for instance, is that there are 5 pennies in the cup.

Later variations of counting events involve children making Ts (for tens) or lines (for ones) each time they count. Like counting the number of coins dropped, the number of Ts or lines can be checked by counting the Ts or lines made.

The first exercise in which children count claps occurs in Lesson 9.

EXERCISE 3: COUNTING EVENTS

Note: Clap at one-second intervals.

a. We're going to count claps. My turn to count the claps. (Clap) 1, (clap) 2, (clap) 3.
• Everybody, how many times did I clap? (Signal.) 3.
b. Let's do it together. Count the claps. (Clap 3 times as you and children say:) *1, 2, 3.*
(Repeat until firm.)
• How many times did I clap? (Signal.) 3.
c. Your turn: Count the claps. (Clap 3.) *1, 2, 3.*
• How many times did I clap? (Signal.) 3.
d. Now I will clap 5 times. Count the claps. (Clap 5.) *1, 2, 3, 4, 5.*
(Repeat until firm.)
• How many times did I clap? (Signal.) 5.

INDIVIDUAL TURNS

(Call on individual children to perform one of the following tasks.)

• Count the claps. (Clap 3.) *1, 2, 3.*
How many times did I clap? *3.*
• Count the claps. (Clap 5.) *1, 2, 3, 4, 5.*
How many times did I clap? *5.*

Lesson 9, Exercise 3

In step A, you model counting claps. Do not say the number at the same time you clap. Rather say the number clearly after you clap. Then pause before your next clap. When children count with you (step B) and count by themselves (step C) use the same timing.

The clap-first convention is important because if the children try to count at the same time you clap, some children will not know when you have clapped for the last time and may count your last clap twice.

In step D, you tell the children that you will clap 5 times. You clap, and children count.

Teaching Note: Stop students as soon as you notice not all of them responding. Repeat the step two times after a mistake.

It's particularly important to practice this exercise before you present it to the children. The reason is that you're not holding the presentation book during the time you're clapping. So the more familiar you are with the script, the smoother your presentation will be.

Be sure to provide individual turns after the group is firm to verify children are able to count claps.

In later exercises, children count other events, such as the number of times the teacher drops coins into a cup. This activity is important because it shows the relationship of counting events with counting objects. For example, after children have counted 6 coins dropping into the cup, the teacher asks how many coins are in the cup. The teacher then pours the coins from the cup. Children count them and confirm that 6 coins were in the cup.

After children have practiced a variety of strategies for counting objects and writing some numbers, children write numbers for lines. The discussion and sample exercises for writing numbers for lines appears in the **Equality and Equations** track (beginning on page 76).

The final application of Counting Events before Counting is integrated into Problem-Solving Operations involves making lines and counting them. For instance, the children make 10 lines. Each line they make is an event they count. After children have made the lines, they can confirm the number by counting the lines.

Here's part of the exercise from Lesson 21:

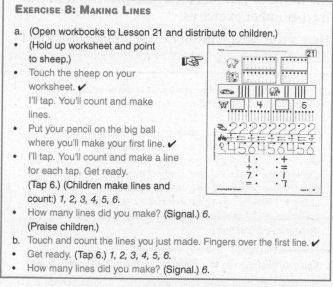

EXERCISE 8: MAKING LINES

a. (Open workbooks to Lesson 21 and distribute to children.)
• (Hold up worksheet and point to sheep.)
• Touch the sheep on your worksheet. ✔
  I'll tap. You'll count and make lines.
• Put your pencil on the big ball where you'll make your first line. ✔
• I'll tap. You'll count and make a line for each tap. Get ready.
  (Tap 6.) (Children make lines and count:) *1, 2, 3, 4, 5, 6.*
• How many lines did you make? (Signal.) *6.*
  (Praise children.)
b. Touch and count the lines you just made. Fingers over the first line. ✔
• Get ready. (Tap 6.) *1, 2, 3, 4, 5, 6.*
• How many lines did you make? (Signal.) *6.*

from Lesson 21, Exercise 8

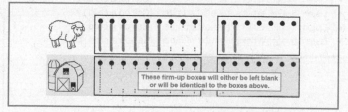

Lesson 21, Answer Key

In step A, the teacher taps 6 times; children count and make a line for each tap. After completing the set of lines, children touch and count the lines to confirm that there are 6 lines (step B).

Teaching Note: Children are to respond verbally as soon as you tap. They are not to make the line and then count. Rather, they say the number as they start to make the line.

Keep your timing for the taps slow enough that children are able to get their pencils in place for making the next line before your next tap. It's important for you to watch children as they make lines.

Most groups will need additional practice on each example. The screened boxes next to the barn are provided for this optional extra practice. If children are not firm on the original trial, have them make "pretend lines" on the screened boxes before they make actual lines. Direct children to put their pencils down and use their finger to touch where they will make the line for each tap. When you tap, they are to say the number as they quickly pretend to make the line. As soon as they finish a pretend line, they are to touch the ball for the next line. After children have attained reasonable performance simultaneously counting and making pretend lines, direct them to do the same task using their pencils.

Praise children for making lines quickly and for immediately going to the ball for the next line.

COUNT ON: NEXT NUMBER

Starting on Lesson 14 children count on by starting with a number other than 1. They get the number they start with going by holding it. For 6, they say *siiix*. For 9 they say *niiine*. Then they count from that number. The reason for teaching children to "get it going" is that the number they get going tells the amount they already have. The next numbers (the numbers that follow) are those that "count on" from the number they get going. The most difficult part of counting on is not to get a number going but to say the correct next number. This is the critical number—the one children most frequently miss. The numbers that follow the first two numbers are easier because they are part of a familiar counting sequence.

Learning to count on is easier if children have learned to identify the next number. We noted earlier that in Lesson 1 (Exercise 7) children are introduced to *next number*. In subsequent lessons, children work with a series of familiar numbers. The easiest series starts with 1. The teacher counts: "1, 2, 3." The teacher gets the last number in the series going, "fouuur." Then the teacher asks, "What's the next number?"

To present a version that is slightly more difficult, the teacher repeats the task above but starts at a number other than 1: "2, 3, fouuur. What's the next number?"

For some tasks, the teacher directs children to say the next number when she points at the children. Like the examples below, the teacher counts a familiar series as she points at herself and gets the last number going: "5, 6, sevennn." Then she points at the children to signal them to say the next number. This version of the next-number task is easier for children because the teacher doesn't interrupt the counting sequence with the question "What's the next number?"

Gradually the next-number tasks become more difficult as children work with shorter series: "6, sevennn, what's the next number?" After children receive ample practice saying the next number as part of a familiar counting series, the most difficult type is presented. The teacher gets a number going, "sevennn," and asks, "What's the next number?"

Note that the introduction of these types is keyed to the amount of practice children have had with the range of numbers that are presented in the next-number exercises.

Here's the Next Number exercise from Lesson 11:

EXERCISE 8: NEXT NUMBER

a. We'll count and end up with 10. What number will we end up with? (Signal.) *10.*
- Count and end up with 10. Get ready. (Tap 10 as you and children count:) *1, 2, 3, 4, 5, 6, 7, 8, 9, 10.*
- (Repeat step a until firm.)

b. All by yourself: Count and end up with 10. Get ready. (Tap 10.) *1, 2, 3, 4, 5, 6, 7, 8, 9, 10.*
- (Repeat step b until firm.)

c. I'm going to count, and you're going to tell me the next number. It will be hard because I won't start with 1 for all the numbers.
- Listen: 6, 7, eieieight. What's the next number? (Signal.) *9.*
- Listen: 2, threee. What's the next number? (Signal.) *4.*
- Listen: 3, fouuur. What's the next number? (Signal.) *5.*
- Listen: 5, 6, sevennn. What's the next number? (Signal.) *8.*
- Listen: 4, fiiive. What's the next number? (Signal.) *6.*
- (Repeat step c until firm.)

d. Now I'll count. You'll tell me the next number when I point at you. For some of the numbers, I won't start at 1.
- (Point to yourself.) Listen: 5, siiix. (Point at children.) *7.*
- (Point to yourself.) Listen: threee. (Point at children.) *4.*
- (Point to yourself.) Listen: 6, sevennn. (Point at children.) *8.*
- (Point to yourself.) Listen: 4, fiiive. (Point at children.) *6.*
- (Point to yourself.) Listen: 6, 7, eieieight. (Point at children.) *9.*
- (Repeat step d until firm.)

e. Let's do it again, but this time you'll answer the question.
- Listen: fiiive. What's the next number? (Signal.) *6.*
   (To correct:)
- Listen: 3, 4, fiiive. What's the next number? (Signal.) *6.*
- Listen: fiiive. What's the next number? (Signal.) *6.*
- Listen: twooo. What's the next number? (Signal.) *3.*
- Listen: 5, siiix. What's the next number? (Signal.) *7.*
- Listen: threee. What's the next number? (Signal.) *4.*
- Listen: 6, 7, eieieight. What's the next number? (Signal.) *9.*
- Listen: wuuun. What's the next number? (Signal.) *2.*
- (Repeat step e until firm.)

==== INDIVIDUAL TURNS ====
(Call on individual children to perform one or two of the following tasks.)

- Listen: 6, 7, eieieight. What's the next number? (Call on a child.) *9.*
- Listen: fiiive. What's the next number? (Call on a child.) *6.*
- Listen: threee. What's the next number? (Call on a child.) *4.*
- Listen: 5, siiix. What's the next number? (Call on a child.) *7.*
- Listen: fouuur. What's the next number? (Call on a child.) *5.*
- Listen: twooo. What's the next number? (Call on a child.) *3.*

Lesson 11, Exercise 8

In step A, the teacher models counting to 10 and counts with the children to 10. After this counting is firmed, children count to 10 by themselves in step B. In step C, the teacher counts familiar series of numbers that do not start with 1 and asks children, "What's the next number?"

In step D, a new counting series (5, siiix) is presented and the series for other numbers are repeated but shortened. In this step, the teacher points at the children but doesn't ask the question "What's the next number?"

Step E presents the most difficult version of the next-number tasks for less than 6. The teacher gets different numbers going and asks, "What's the next number?"

In step E, the script specifies a correction. See Teaching Note for more information about the correction. If children make the same kind of mistake for any of the other tasks in step E, use the same correction procedure.

Individual turns are specified after step E. Each number the teacher gets going for Individual Turns was presented in earlier steps of the exercise. The Individual Turns that are specified for numbers less than 6 are the most difficult version of the next-number tasks.

Teaching Note: Make sure children are firm on counting to 10 (step B) before you move on to step C. Children will have difficulty with the rest of the exercise if they are not firm on counting to 10.

This correction shows the procedure for correcting the first task in step E.

The practice for counting should focus on the numbers after 6. For several lessons, children have been performing a multitude of counting tasks with numbers less than or equal to 6, so they should be very firm with the counting series up to 6.

After step B, make sure you hold the number that comes immediately before children respond. Holding the number calls children's attention to the number, and it gives them think-time. This convention will later be expanded when children use the count-on strategy.

After presenting an easier version of a task to elicit the correct response, repeat the task the program originally presented to make sure that children are able to perform it. Use the strategies above to correct each mistake. If children make more than one or two mistakes in a step, repeat the step. For example, if children make two mistakes in step E, the teacher corrects each mistake and continues to the end of step E. Then, the teacher repeats step E again before presenting Individual Turns.

Correction:

For the tasks in steps C, D, and E, if children are not able to say the correct next number, tell them the answer and repeat the task:

Teacher: Listen: 6, 7, eieieight. What's the next number?

Some children: 8.

Teacher: 9. What number?

Children: 9.

Teacher: Listen again: 6, 7, eieieight. What's the next number?

If mistakes persist, make the task easier. For example, if children respond to the last task presented above with the answer "10," make the task easier by presenting a longer series of familiar numbers.

Teacher: 3, 4, 5, 6, 7, eieieight. What's the next number?

If children continue to have difficulty, make the tasks in steps C and E even easier by eliminating the question.

Teacher: Listen again: (Point at yourself.) 3, 4, 5, 6, 7, eieieight. (Point to children.)

COUNT ON: GET IT GOING AND SAY THE NEXT NUMBER

By Lesson 14, children get a number going and count on from that number. Here's the exercise from Lesson 15:

EXERCISE 8: GET IT GOING

a. We're going to get numbers going and say the next number.
- My turn to get 6 going and say the next number. Siiix. (Signal.) 7.
- Your turn: Get 6 going. Siiix. Next number. (Signal.) 7.

b. My turn to get 4 going and say the next number. Fouuur. (Signal.) 5.
- Your turn: Get 4 going. *Fouuur.* Next number. (Signal.) *5.*
- (Repeat step b until firm.)

c. My turn to get 2 going and say the next number. Twooo. (Signal.) 3.
- Your turn: Get 2 going. *Twooo.* Next number. (Signal.) *3.*

d. My turn to get 7 going and say the next number. Sevennn. (Signal.) 8.
- Your turn: Get 7 going. *Sevennn.* Next number. (Signal.) *8.*

e. My turn to get 5 going and say the next number. Fiiive. (Signal.) 6.
- Your turn: Get 5 going. *Fiiive.* Next number. (Signal.) *6.*
- (Repeat steps b through e until firm.)

f. This time we're going to get numbers going and count.
- My turn to get 7 going and count. Sevennn. (Tap 3.) 8, 9, 10.
- Your turn: Get 7 going. *Sevennn.* Count. (Tap 3.) *8, 9, 10.*

g. My turn to get 5 going and count. Fiiive. (Tap 3.) 6, 7, 8.
- Your turn: Get 5 going. *Fiiive.* Count. (Tap 3.) *6, 7, 8.*
- (Repeat steps f and g until firm.)

Lesson 15, Exercise 8

In steps A through E, children get numbers going and say the next number. In steps F and G, children get numbers going and count on. The teacher models each task before directing children to perform it.

Teaching Note: Timing is very important for this exercise. You need to establish a very predictable rhythm. When you model each task, hold the number you get going for more than a second before you say the next number. For steps F and G, after you hold the number you get going, count the following numbers at the rhythmical rate you've established for other counting tasks. Make sure that when children do the task you have just modeled, they use the same timing. Holding the number you get going for more than a second will make later next-number and counting-on tasks easier. The longer duration for the number you get going makes it easier for children to respond in unison and gives children more time to think about the next number.

This exercise is very important to practice because variations of the tasks are used throughout the program.

In Lesson 14, children apply the count-on strategy to counting lines. Before this time, children have counted single groups of lines. This exercise requires children to count two groups of lines.

EXERCISE 8: COUNTING SEPARATE GROUPS

a. (Display page.)   [14:8A]
   Here are two groups of lines.
   (Point to IIII.) Here's one group.
   (Point to IIIII.) Here's another group. You're going to count the lines in each group.

• (Point to IIII.) Count the lines in this group. Get ready. (Touch each line.) *1, 2, 3, 4.*
• How many lines are in this group? (Touch.) *4.*
   Yes, 4.

b. (Point to IIIII.) Count the lines in this group. Get ready. (Touch each line.) *1, 2, 3, 4, 5.*

• How many lines are in this group? (Touch.) *5.*
   Yes, 5.

c. I'm going to count the lines in **both** groups. (Point to IIII.) I'll count the lines in this group. Then I'll just keep on counting. Watch. (Touch lines in first group.) 1, 2, 3, fouuur. (Touch lines in second group.) 5, 6, 7, 8, 9.

d. Do it with me. Get ready. (Touch lines in first group.) *1, 2, 3, fouuur.* (Touch lines in second group.) *5, 6, 7, 8, 9.*
   (Repeat step d until firm.]

e. This time I'll do the first group. Then you'll keep on counting. (Touch lines in first group.) 1, 2, 3, fouuur. (Touch lines in second group.) *5, 6, 7, 8, 9.*
   (Repeat step e until firm.)

f. How many lines in both groups? (Signal.) *9.*

Lesson 14, Exercise 8

In steps A and B, children count the lines in each group. In step C, the teacher models by touching and counting lines in the first group, 1, 2, 3, then holding 4; then touching and counting lines in the second group, 5, 6, 7, 8, 9.

In step D, the teacher directs children to do the counting with her as she repeats the touching and counting. In step E, the teacher touches and counts the lines in the first group, holding fouuur. Then the teacher touches lines in the second group as the children count 5, 6, 7, 8, 9. The bracket shows that steps D and E are to be repeated until children are firm.

Teaching Note: For the final bullet in steps A and B, touch the last line in the group (the fourth and fifth line respectively).

For steps D and E, children are firm if they hold 4 as long as you are touching the last line in the first group and if they clearly say the next numbers as you touch the lines in the second group. Be very strict about children saying the numbers when you touch the lines. If children lag or count ahead, stop and tell them, "Count when I touch the line." And repeat the task.

If children do well the first time you present step E, tell them "good counting" and repeat the step two more times.

On the following lessons, children are responsible for initiating more details of counting two groups of lines. Following is the exercise from Lesson 20 (Exercise 8):

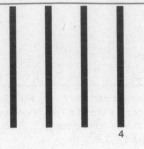

4

EXERCISE 8: COUNTING TWO GROUPS

a. (Display page and point to lines.) [20:8A]
Here are 2 groups of lines. I'll count the lines for the first group. You'll keep counting for the rest of the lines. (Touch lines.) 1, 2, 3, 4.
• How many lines in the first group? (Touch.) *4.*
• (Touch line **4.**) Now you'll get it going and count the rest of the lines. Get 4 going. *Fouuur.* (Touch lines in second group.) *5, 6, 7, 8, 9.*
• How many lines in both groups? (Touch.) *9.*

b. There are 4 lines in the first group.
• Tell me to touch line 4. (Signal.) *Touch line 4.*
• (Touch line **4.**) Which line am I touching? (Tap.) *4.*

⎡ c. (Keep touching line **4.**) You'll do the hard part again. Get 4 going.
⎢ *Fouuur.* (Touch lines in second group.) *5, 6, 7, 8, 9.*
⎣ (Repeat until firm.)
• How many lines in both groups? (Touch.) *9*

d. (Open workbooks to Lesson 20 and distribute to children.)
• (Hold up worksheet and point to pencil.) ☞
• Touch the pencil on your worksheet. ✔
The lines next to the pencil are the same groups of lines you just counted. There are 4 lines in the first group.
⎡ • Fingers up. ✔
⎢ • Everybody, touch line 4. ✔
⎢ • Again, touch line 4 and keep touching it. ✔
⎣ (Repeat until firm.)
⎡ e. I'll get 4 going. You'll touch and count the rest of the lines. *Fouuur.* Touch and count. (Tap 5 as children touch and count.) *5, 6, 7, 8, 9.*
⎣ (Repeat step e until firm.)

⎡ f. This time you'll get 4 going and touch and count the rest of the lines.
⎢ • Touch line 4. ✔
⎢ • Get it going. *Fouuur.* Touch and count. (Tap 5.) *5, 6, 7, 8, 9.*
⎣ (Repeat until firm.)
• How many lines are in both groups? (Signal.) *9.*
g. Touch the cat. ✔
There are 2 groups of lines next to the cat. I'll tap. You'll touch and count the lines in the first group.
• Fingers over the first group. ✔
• Touch and count. Get ready. (Tap 6.) *1, 2, 3, 4, 5, 6.*
• How many lines are in the first group? (Signal.) *6.*

⎡ h. Fingers up. ✔
⎢ • Everybody, touch line 6. ✔
⎢ • Fingers up. ✔
⎢ • Again: Touch line 6 and keep touching it. ✔
⎣ (Repeat step h until firm.)
⎡ i. You'll touch and count the rest of the lines.
⎢ • Get 6 going. *Siiix.* Touch and count. (Tap 2.) *7, 8.*
⎣ (Repeat step i until firm.)
j. Let's do it again. Touch line 6. ✔
• Get 6 going. *Siiix.* Touch and count. (Tap 2.) *7, 8.*
• How many lines are in both groups? (Signal.) *8.*
Good counting lines.

Lesson 20, Exercise 8

In step A, children count the lines in two groups. They count the four lines in the first group. Then the teacher touches the fourth line. Children get 4 going and count on for the lines in the second group.

In step B, the children tell the teacher to touch line 4. Children "get 4 going" as the teacher touches the last line in the first group. Then they count the remaining lines as the teacher touches them.

In step C, children repeat the counting as the teacher touches the lines. Then they say the number of lines in both groups.

In steps D through F, children count on for the same two groups of lines in their workbook. (These groups appear next to a picture of a pencil.)

In steps G through J, children do the same counting operation with a new pair of groups (next to a picture of a cat).

Teaching Note: In steps A through C, make sure that children hold *fouuur* as long as you touch the last line in the first group. Then make sure children count for each line as you touch it. Throughout the program, children will use the convention of getting the number for the last line going and count on for the remaining lines (even when no lines, only numbers, are shown).

Steps D through F assure that children coordinate touching and counting as they get the number going, then touching and counting for the lines in the second group.

A good plan is to repeat steps E and F a few times even if children are not making mistakes, so you can confirm that children are touching in time with their counting.

If children make a counting error or a touching error, stop them immediately. Model the touching and timing that children had difficulty with. Then repeat the step.

Steps G though J are very important. Children are touching and counting for new groups of lines without following the teacher's model first. Children's performance on these steps is a good indication about their mastery of the count-on skills that have been presented. If children make more than one mistake on steps G though J, repeat Exercise 8 steps D through J.

COUNTING FROM-TO

During the early lessons, children receive practice in counting from numbers other than 1. The teacher models "the hard parts" of sequences children are learning. Then children repeat the hard part. For instance:

- Listen to the hard part: 9, 10, 11. Say the hard part with me: **9, 10, 11.**
- Your turn: Say the hard part.

On Lesson 20, children are introduced to getting a number going and counting to a specified number.

Here's the exercise:

EXERCISE 3: ROTE COUNTING

a. My turn to get 12 going and count to 15. Tweeelve, (Tap 3.) 13, 14, 15.
- Your turn to get 12 going and count to 15. Get it going. *Tweeelve.* Count. (Tap 3.) *13, 14, 15.*
 (Repeat step a until firm.)

b. Now you're going to start with 10 and count to 15.
- What number are you going to start with? (Signal.) 10.
- What number are you going to end up with? (Signal.) 15.
- Get 10 going. *Tennn.* Count. (Tap 5.) *11, 12, 13, 14, 15.*
- Once more: Get 10 going. *Tennn.* Count. (Tap 5.) *11, 12, 13, 14, 15.*
 (Repeat step b until firm.)

c. Now you'll start with 9 and count to 13.
- What number are you going to start with? (Signal.) 9.
- What number are you going to end up with? (Signal.) 13.
- Get 9 going. *Niiine.* Count. (Tap 4.) *10, 11, 12, 13.*
 (Repeat step c until firm.)

d. Start with 1 and count to 15. Get ready. (Tap 15.) *1, 2, 3, 4, 5, 6, 7, 8, 9, 10, 11, 12, 13, 14, 15.*
 (Repeat step d until firm.)

==================== INDIVIDUAL TURNS ====================
(Call on individual children to perform one of the following tasks)

- Get 10 going and count to 15. *Tennn, 11, 12, 13, 14, 15.*
- Get 9 going and count to 13. *Niiine, 10, 11, 12, 13.*
==

Lesson 20, Exercise 3

In step A, the teacher models getting 12 going and counting to 15. Then children get 12 going and count to 15. The ending number for all tasks in this exercise is 15.

In steps B and C, children do not receive a model—simply information about the starting number and ending number. ("What number are you going to start with?...What number are you going to end up with?...") The starting number in step B is 10. The ending number is 15.

In step D, children count from 1 to 15. Children do not get the first number going (1). They just count after each tap.

Individual Turns (specified after step D) are critical for this exercise because it is very easy not to hear the specific mistakes children may make in their group responses. Sample possibly two children for each task specified. (See the section **Individual Turns** in the previous section, Teaching Effectively, for information about procedures and strategies for giving Individual Turns.)

The tasks in steps A through C are different from the earlier exercises that presented "the hard parts" because children get the first number going, then count.

The following lessons provide regular practice in counting from a number to a number. By Lesson 50, children should reliably count from-to and stop at the correct number. If children still count past a number after Lesson 50, be more vigilant about correcting these errors, because in the 60s lesson range, children learn to solve problems of the following type:

$$2 + \boxed{} = 9$$

The strategy children learn is to get the first number going and count to the number after the equals as they make a line under the box for each number they count (then they count the lines and write the number in the box to complete the equation). For this problem type, the end number is shown, so it serves as a prompt about when children should stop counting. However, children may become distracted because they make lines as they count.

COUNTING BY TENS

Children first count by tens on Lesson 49. The skill is later used in counting for Ts. Each T stands for ten. Children learn to express two-digit numbers as Ts and lines. For instance, 25 is 20 plus 5, which is 2 Ts plus 5 lines:

T T I I I I I

Children also count by tens when they count coins:

To compute the number of cents in the group of coins, children count by tens for each dime, then continue counting by ones for each penny (10, 20, 30, 31, 32, 33, 34).

Finally, children count by tens for all two-digit column problems—addition and subtraction.

Here's the exercise from Lesson 49 that introduces Counting by Tens:

EXERCISE 2: COUNTING BY TENS

a. You know how to count by ones.
- When you count by ones, the first number you say is 1. When you count by ones, what's the first number? (Signal.) *1.*
- Listen: When you count by tens, the first number you say is 10. What's the first number you say when you count by tens? (Signal.) *10.*

b. My turn to count by tens to 50: 10, 20, 30, 40, 50.
- Do it with me. Count by tens to 50. Get ready. (Tap 5.) *10, 20, 30, 40, 50.*
 (Repeat step b until firm.)

c. All by yourself: Count by tens to 50. Get ready. (Tap 5.) *10, 20, 30, 40, 50.*
 (Repeat step c until firm.)

d. Now, I'll say a tens number. Then you'll tell me the next number for counting by tens.
- Listen: 10. What's the next tens number? (Signal.) *20.*
- Yes, 20. What's the next tens number? (Signal.) *30.*
- Yes, 30. What's the next tens number? (Signal.) *40.*
- Yes, 40. What's the next tens number? (Signal.) *50.*

━━━━━━ **INDIVIDUAL TURNS** ━━━━━━

(Call on individual children to perform one of the following tasks.)

 Tell me the next number for counting by tens.
- Listen: 10. What's the next tens number? (Call on a child.) *20.*
- 20. What's the next tens number? (Call on a child.) *30.*
- 30. What's the next tens number? (Call on a child.) *40.*
- 40. What's the next tens number? (Call on a child.) *50.*

Lesson 49, Exercise 2

In step B, the teacher models counting by tens to 50. Then the teacher and children count together. In step C, children count by themselves. In step D, children indicate the next tens number in the sequence. After step D, individual children are called on to identify the next tens number in the sequence.

Teaching Note: By Lesson 49, children are well practiced counting by ones to __.

Most of the sequence for counting by tens has a very similar sound to the sequence for counting teen numbers; 13, 14, 15, 16...sounds very similar to 30, 40, 50, 60. Children say that part of the sequence quickly. However, some children may say the teen numbers instead of saying tens numbers. Listen closely during steps B through D to make sure children are saying, "thirty, forty, fifty" not "thirteen, fourteen, fifteen." Use Individual Turns to make sure they are pronouncing these tens numbers correctly.

Present, correct, and firm steps B and C, using the same procedures and strategies used to firm rote counting by ones.

If children have trouble indicating the next number in step D, model the sequence with the questions but without confirmation:

> My turn: Ten. What's the next tens number? 20.
>
> What's the next tens number? 30.
>
> What's the next tens number? 40.
>
> What's the next tens number? 50.

Then repeat step D and the Individual Turns at the same pace you modeled.

If a child makes a mistake, correct the group. Then repeat the Individual Turns from the beginning. Starting the Individual Turns over for this exercise is important because children are not yet expected to identify the next tens number out of sequence.

As part of what children learn, they practice identifying the next tens number out of sequence. By Lesson 60, children should be fluent in counting by tens to one hundred and identifying the next tens number.

Here's the exercise from Lesson 59:

EXERCISE 1: ADDING TENS

a. Count by tens to one hundred. Get ready. (Tap 10.) *10, 20, 30, 40, 50, 60, 70, 80, 90, 100.*

b. When you count by tens, what's the first number you say? (Signal.) *10.*
- What's the next tens number? (Signal.) *20.* Yes, 20.
- What's the next tens number? (Signal.) *30.*
- (Repeat for *40, 50, 60, 70, 80, 90.*)
- (Repeat step b until firm.)

c. Now I'll mix them up.
- Listen: 70. What's the next tens number? (Signal.) *80.*
- New number: 90. What's the next tens number? (Signal.) *100.*
- New number: 20. What's the next tens number? (Signal.) *30.*
- New number: 80. What's the next tens number? (Signal.) *90.*
- (Repeat step c until firm.)

Lesson 59, Exercise 1

In step A, children count by tens to one hundred. In step B, children say the next tens numbers in the sequence. In step C, children say some of the next tens numbers that are not presented in the sequence.

After children practice counting by tens to 50, the program systematically extends the counting to later tens numbers.

The table below shows the lesson number at which children learn to count by tens to different decade numbers.

Counting by Tens

Beginning Lesson	49	52	53	53	55	57
Children count by tens to	50	60	70	80	90	100

Teaching Note: Children should not have trouble with any of the tasks in steps A and B. The tasks that they will most likely have trouble with are the second and third bullets in step C. For the first and fourth bullets, children say the next tens number for seventy and eighty, respectively. This task is exactly the same as the next-number task for 7 and for 8 except that the syllable *ty* is added to the end of each number. The second and third bullets do not follow this regular pattern. For the second bullet, the number that comes after ninety is not "tenty," it's one hundred.

If children have trouble with any of the tasks in step C, use parallel correction procedures and prompting strategies that you would use when correcting for saying the next-number counting by ones. Tell children the answer and repeat the question. Present a simplified version of the same task. Then present the task in its original form.

For example, if children said the incorrect response to the task for the second bullet of step C:

Teacher: New number: 90. What's the next tens number?

Some children: *10.*

Teacher: One hundred. What tens number?

Children: *100.*

Teacher: Listen: 60, 70, 80, ninetyyy. What's the next tens number?

Children: *100.*

Teacher: Yes. Listen again: 90. What's the next tens number?

Then, repeat step C from the beginning.

Note that the teacher models *one* hundred. It's important children understand that the first part is *one*.

If children made several mistakes on this exercise, provide a delayed test on the entire exercise by presenting it again after presenting Exercise 2, 3, and maybe 4. (See the **Firming Delayed Tests** on page 36 of this guide for additional information.)

TS AND NUMBERS

Ts are introduced on Lesson 60.
Here's the exercise from Lesson 60:

EXERCISE 6: COUNTERS FOR TENS

a. (Display page and point to **Ts**.)   [60:6A]
- These are groups of Ts. What are they? (Touch.) *Groups of Ts.*
b. Listen: You count by tens to get a number for Ts.
  What do you count by to get a number for Ts? (Signal.) *Tens.*
- What do you count by to get a number for lines? (Signal.) *Ones.*
c. (Point to **T T T T T.**) There are 5 Ts in this group.
  To figure out how many are in this group, I count by ten for each T.
  Watch. (Touch each T.) 10, 20, 30, 40, 50.
- What's the number for this group? (Touch.) *50.*
d. (Point to **T T.**) Look at the Ts in this group.
- How many Ts are in this group? (Touch.) *2.*
- What do I count by for each T? (Signal.) *10.*
  (Touch each T.) 10, 20.
- What's the number for this group? (Touch.) *20.*

e. What do you count by for each T? (Signal.) *10.*
- (Point to **T T T T T.**) Your turn to count. Get ready. (Touch each T.)
  10, 20, 30, 40, 50.
- What's the number for this group? (Touch.) *50.*
f. (Point to **T T.**) Count for the Ts in this row. Get ready. (Touch each T.) *10, 20.*
- What's the number for this group? (Touch.) *20.*
g. (Point to **T T T T T T T T.**) Count for the Ts in this group. Get ready. (Touch each T.) *10, 20, 30, 40, 50, 60, 70, 80.*
- What's the number for this group? (Touch.) *80.*
  (Repeat steps e through g that were not firm.)

In step B, children learn that when they count for Ts they count by tens.

Children apply this rule to the different examples in steps C through G.

> **Teaching Note:** If children are fluent at counting by tens, they should not have difficulty learning to count groups of Ts. The skills for touching, counting, and identifying the number for a group of Ts are parallel to the skills used for counting by ones for lines.

After working on identifying the numbers for groups of Ts, children discriminate counting for groups of Ts or counting for groups of lines. Children identify whether they count by ones or by tens. Then they count.

In Lesson 72, children learn that the name of most tens numbers tells the number of Ts to write.

> The name for 70 is seven-T. You write 7 Ts.
> The name of 40 is four-T. You write 4 Ts.

Also, in Lesson 72, children are introduced to groups that have both Ts and lines.

Here's a part of the exercise:

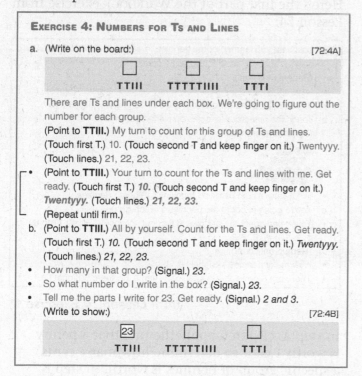

from Lesson 72, Exercise 4

In step A, the teacher models the counting and leads children in counting. In step B, children count without the teacher. Then they answer questions about the number for the group (23). Children follow the same procedure in the remaining part of the exercise (not shown here) to figure out the numbers for 54 and 31.

> **Teaching Note:** You use the same strategy that you use earlier for counting on. You hold the last number for Ts, then count on for the lines. Use the same pacing and the same basic corrections specified for counting on. Keep touching the last T and hold the last number for the Ts about two seconds. Then touch and count at the regular pace for the lines. Holding the last number for the Ts not only provides children with think-time, it also alerts them to a change in what they count by.
>
> It is important for children to count the correct number for each counter. It is also important that they coordinate their counting with the teacher's touching. Remember to keep the same pacing for the tasks in this exercise and for exercises of the same type in subsequent lessons.

On the following lessons, children become increasingly responsible for counting for Ts and lines without teacher support.

Starting in Lesson 73, children touch and count for groups shown in their workbook.

Following are two groups children analyze:

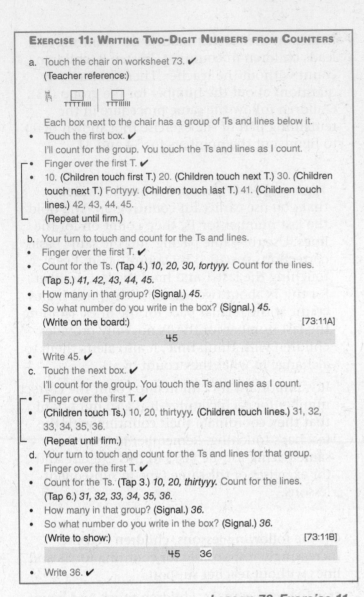

EXERCISE 11: WRITING TWO-DIGIT NUMBERS FROM COUNTERS

a. Touch the chair on worksheet 73. ✔
   (Teacher reference:)

   Each box next to the chair has a group of Ts and lines below it.
   • Touch the first box. ✔
     I'll count for the group. You touch the Ts and lines as I count.
   • Finger over the first T. ✔
   • 10. (Children touch first T.) 20. (Children touch next T.) 30. (Children touch next T.) Fortyyy. (Children touch last T.) 41. (Children touch lines.) 42, 43, 44, 45.
     (Repeat until firm.)

b. Your turn to touch and count for the Ts and lines.
   • Finger over the first T. ✔
   • Count for the Ts. (Tap 4.) *10, 20, 30, fortyyy.* Count for the lines. (Tap 5.) *41, 42, 43, 44, 45.*
   • How many in that group? (Signal.) *45.*
   • So what number do you write in the box? (Signal.) *45.*
     (Write on the board:)   [73:11A]

45

   • Write 45. ✔

c. Touch the next box. ✔
   I'll count for the group. You touch the Ts and lines as I count.
   • Finger over the first T. ✔
   • (Children touch Ts.) 10, 20, thirtyyy. (Children touch lines.) 31, 32, 33, 34, 35, 36.
     (Repeat until firm.)

d. Your turn to touch and count for the Ts and lines for that group.
   • Finger over the first T. ✔
   • Count for the Ts. (Tap 3.) *10, 20, thirtyyy.* Count for the lines. (Tap 6.) *31, 32, 33, 34, 35, 36.*
   • How many in that group? (Signal.) *36.*
   • So what number do you write in the box? (Signal.) *36.*
     (Write to show:)   [73:11B]

45	36

   • Write 36. ✔

Lesson 73, Exercise 11

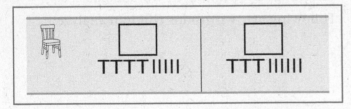

Lesson 73, Workbook

In step A, the teacher counts and the children touch the Ts and lines for the first group. In step B, children touch and count for the first group and identify the number that goes in the box. *Note:* The teacher paces the counting with taps. Finally, the teacher writes that number on the board (or displays it), and children copy it in their workbook box. Steps C and D are parallel to steps A and B for the second group of counters.

Teaching Note: Repeat the tasks when you and children count for the Ts and lines. Watch children's responses. Make sure they are touching the last T until you start counting on for the lines. If children have trouble, stop them immediately and correct.

Use the same pacing in step B (first two bullets) when children touch and count. Tap for each response children are to make. Tap loud enough to assure that children hear the audio timing.

For this exercise, it is also a good idea to repeat the first two bullets in steps B and D to observe that children are touching and counting correctly.

COUNTING FOR COINS

In *CMC Level A*, children learn to count cents for pennies, nickels, dimes, and quarters. Children apply counting by ones and by tens to pennies and dimes.

In Lesson 71 (Exercise 6) children are introduced to pennies. In Lesson 74, children learn that a penny is worth 1 cent.

Here's the first part of the Workbook exercise from Lesson 74:

a. Last time you learned about pennies. A penny is worth 1 cent. How much is a penny worth? (Signal.) *1 cent.*
   (Repeat until firm.)
   • Listen: If I have 6 pennies, how many cents do I have? (Signal.) *6.*
   • If I have 134 pennies, how many cents do I have? (Signal.) *134.*
b. Touch the first group of coins on worksheet 74. ✔
   (Teacher reference:)

   • What kind of coins are you touching? (Signal.) *Pennies.*
c. You're going to count the pennies in each group.
   • Touch and count the pennies in the first group. Get ready. (Tap 8.)
     1, 2, 3, 4, 5, 6, 7, 8.
     (Repeat until firm.)
   • How many pennies? (Signal.) *8.*
   • There are 8 pennies. So how many cents are there? (Signal.) *8.*

from Lesson 74, Exercise 11

In step A, children apply the rule that a penny is worth 1 cent to determine how many cents a specific group of pennies is worth. In step B, children indicate the kinds of coins that are in the groups—pennies. In step C, children touch and count the first group of pennies and then indicate the number of cents for the group.

In the steps that follow (but aren't shown here), children use the same procedure for identifying the cents for the other group. Children write the number (step D) after the equals signs (step E), = 8 and = 11, respectively.

pennies that are in the second row, hold up a workbook and model the touching and counting.

Children learn about dimes on Lesson 81. They learn that each dime is worth 10 cents. To figure out how much a group of dimes is worth, children count by ten for each dime.

Here's the exercise from Lesson 82:

EXERCISE 5: COINS—*Counting Cents*

a. (Display page and point to coins.) [82:5A]
We're going to figure out the number of cents in this group.
(Point to first dime.) Is this a nickel, a dime, or a penny? (Touch.) *A dime.*
- What is a dime worth? (Signal.) *10 cents.*
b. Counting dimes is just like counting Ts. You count by ten for each dime and count by one for each penny. My turn to count the number of cents. (Touch dimes.) 10, 20, thirtyyy. (Touch pennies.) 31, 32, 33, 34, 35.
- Your turn: Count the cents for this group of coins. Get ready. (Touch dimes.) *10, 20, thirtyyy.* (Touch pennies.) *31, 32, 33, 34, 35.*
  (Repeat until firm.)
- How many cents is this group worth? (Signal.) *35.*
c. Today's lesson is 82. What number? (Signal.) *82.*
- Tell me the parts of 82. (Signal.) *8 and 2.*
- (Distribute unopened workbooks to children.) Open your workbook to Lesson 82.
  (Observe children and give feedback.)

d. Touch the groups of coins on worksheet 82. ✔
There are dimes and pennies in each group. You're going to figure out the number of cents for each group.
- What number do you count by for each dime? (Signal.) *Ten.*
- What number do you count by for each penny? (Signal.) *One.*
  (Repeat until firm.)
e. Finger over the first dime. ✔
- Touch and count for the dimes. Get ready. (Tap 2.) *10, twentyyy.*
  Count for the pennies. (Tap 4.) *21, 22, 23, 24.*
  (Repeat until firm.)
- How many cents is that group worth? (Signal.) *24.*
  Later you'll complete the dotted equals and write 24.

Lesson 82, Exercise 5

In step A, the teacher displays a group of coins consisting of 3 dimes and 5 pennies. Children identify a dime and indicate how many cents a dime is worth. In step B, the teacher models the touching and counting for the group of coins. The teacher touches and counts by tens for the 3 dimes, holding thirtyyy, then touches and counts for the pennies: 31, 32, 33, 34, 35. Then the teacher directs the children to count the cents for the coins.

In addition to pennies and dimes, the program teaches nickels (Lesson 77), quarters (Lesson 89), and bills (Lesson 102).

COUNTING AS PLUSING

The sequence begins on 14 and continues through 125. Once children learn to count by ones and count by tens, they receive practice counting orally on almost all lessons. This practice is not designed to teach anything new, it simply provides enough practice that the counting operations become automatic and extensions of the counting sequences become obvious.

"Plusing" is presented as a simple extension of counting on. Early in the program, children start with a number and count by ones, for example, "Start with 7 and count to 10."

Later in the program, the directions refer to "plus" so that children get the idea that counting is simply repeated addition of 1, for example, "Start with 87 and plus ones to 91."

The same relationship is shown for counting by tens. Soon after counting by tens is introduced, children are directed to start with a tens number and count by tens. Later, the tasks refer to plusing tens, for instance, "Start with 37 and plus tens to 67." Once children are well practiced in plusing ones and tens, they work on exercises that require them to plus ones and plus tens.

Counting backward from 5 is introduced on Lesson 88. By Lesson 94, children are able to count backward from 8 without looking at a number line. The same process used for teaching counting by ones and counting by tens is used for teaching counting backward. Counting backward is introduced 88 lessons after children begin work on counting forward. The introduction to counting backward is delayed this long to ensure that children are virtually automatic at counting forward. Also, new numbers to the counting-backward sequences are introduced at a slower rate than new numbers to sequences for counting forward. The slower introduction provides children with more practice.

Here's part of the exercise from Lesson 94 in which children count backward from 8 without a number line and then identify the next number counting backward.

EXERCISE 2: COUNTING BACKWARD

a. I'm going to start with 8 and count backward to 1: 8, 7, 6, 5, 4, 3, 2, 1.
- Your turn: Start with 8 and count backward to 1. Get ready. (Tap 8.) *8, 7, 6, 5, 4, 3, 2, 1.*
  (Repeat step a until firm.)

b. Now I'm going to count backward and stop. When I point at you, tell me the next number when you count backward.
- (Point to yourself.) Listen: 8, 7, 6, 5, fouuur. (Point at children.) *3.*
- (Point to yourself.) New problem: 7, 6, fiiive. (Point at children.) *4.*
- (Point to yourself.) New problem: 5, 4, threee. (Point at children.) *2.*
- (Point to yourself.) New problem: 8, sevennn. (Point at children.) *6.*
  (Repeat step b until firm.)

from Lesson 94, Exercise 2

In step A, the teacher models counting backward from 8 and then children count backward from 8. Step A is repeated until children are firm counting backward from 8. In step B, the teacher tells children that when they are pointed to, they should say the next number for counting backward. The teacher counts familiar series of numbers, gets the last number going, points at children, and they say the next number.

Teaching Note: The same presentation, correction, and firming strategies used for counting-on should be used to present, correct, and firm counting-backward skills.

Here's part of an exercise from Lesson 101 in which children link counting backward to taking away 1:

$$1 \quad 2 \quad 3 \quad 4 \quad 5 \quad 6 \quad 7 \quad 8 \quad 9 \quad 10$$

EXERCISE 5: TAKE AWAY 1

a. Everybody, start with 10 and count backward to 1. Get ready. (Tap while you and children count:) *10, 9, 8, 7, 6, 5, 4, 3, 2, 1.* (Repeat step a until firm.)

b. (Display page and point to number line.)   [101:5A]
   Listen: When you take away 1, the answer is the next number you say when you **count backward.**
   My turn: (Point to **2.**) What's 2 take away 1? *(Touch 1.)* 1.

c. Your turn: (Point to **2.**) What's 2 take away 1? (Touch 1.) *1.*
   Yes, 1.
   • Say the equation for 2 take away 1. Get ready. (Tap 5.) *2 take away 1 equals 1.*
   (Repeat step c until firm.)

d. I'm going to get tricky now.
   • (Point to **5.**) What number is this? (Touch.) *5.*
   What's 5 take away 1? (Touch 4.) *4.*
   (Keep touching 4.) Yes, 4.
   • Say the equation for 5 take away 1. (Tap 5.) *5 take away 1 equals 4.*

e. (Point to **7.**) What number is this? (Touch.) *7.*
   • What's 7 take away 1? (Touch 6.) *6.*
   (Keep touching 6.) Yes, 6.
   • Say the equation for 7 take away 1. (Tap 5.) *7 take away 1 equals 6.*

from Lesson 101, Exercise 5

Note: Children do not count backward when they work problems like 25 – 14. Rather they make counters for 25 and then take away 14 by crossing out one T and 4 lines. Then children count for the remaining Ts and lines.

ORDINAL NUMBERS

Ordinal numbers are introduced on Lesson 89.
On Lesson 89, children learn the names *first*,
second, and *third*.

Here's the exercise:

EXERCISE 6: ORDINAL NUMBERS

a. (Display page and point to dog.)   [89:6A]
- Count the things in this row. Get ready. (Touch.) *1, 2, 3, 4, 5.*
- How many things are in this row? (Signal.) *5.*
- (Point to dog.) This is the first thing. What is the first thing in this row? (Touch.) *(A) dog.*
- Say **first**. (Signal.) *First.*
b. (Point to cat.) This is the **second** thing. What is the second thing in this row? (Touch.) *(A) cat.*
- Say **second**. (Signal.) *Second.*
c. (Point to boy.) This is the **third** thing. What is the third thing in this row? (Touch.) *(A) boy.*
- Say **third**. (Signal.) *Third.*

d. (Point to dog.) My turn: (Touch dog.) First. (Touch cat.) Second. (Touch boy.) Third.
- Say that with me. (Touch objects.) *First, second, third.*
(Repeat step d until firm.)
e. What's the first thing in the row? (Signal.) *(A) dog.*
- What's the second thing in the row? (Signal.) *(A) cat.*
- What's the third thing in the row? (Signal.) *(A) boy.*
(Repeat step e until firm.)

Lesson 89, Exercise 6

On the following lessons, children work with similar displays. Children identify whether specific objects are first, second, or third (as they do in the exercise above) and also, fourth, or fifth. Children continue to name objects when they are referred to by their ordinal position. Example: "What's the fifth thing in this row?"

After children learn first through fifth, they learn that other ordinal numbers (6th–12th) follow a pattern—the ordinal name is the number name followed by the sound *th*. After children have learned ordinal counting from first through fifth, the teacher regularly uses ordinal numbers in directions. These directions use ordinal numbers through *twelfth*. Example: "Touch the 12th problem."

Symbols—Lessons 1–120

Like counting, identifying and writing symbols begins on Lesson 1. Work on symbols continues on every lesson throughout *CMC Level A*. The symbols track develops in three directions: Children learn to read symbols, write symbols, and use symbols to show mathematical relationships.

The order of introduction starts with reading, followed by writing.

The table shows the order of introduction for different symbols and the lesson number on which each symbol (1 through 9, 0, =, +, −, I, and T) is introduced and first written.

Symbol Introduction

Category	Symbol	First Identified on		First Written on Workbook
		Lesson	Exercise	Lesson
single digits	**0**	29	6	31
	1	21	2	27
	2	3	2	4
	3	17	3	23
	4	1	2	1
	5	13	2	15
	6	7	1	8
	7	11	2	12
	8	32	5	38/39
	9	25	1	30
easy -ty numbers ①	**41–49**	52	5	56
	61–69	58	4	65
	71–79	61	5	64
	81–89	61	5	63
	91–99	61	5	62
hard -ty numbers ②	**21–29**	52	5	54
	31–39	53	7	55
	51–59	62	6	66
decade numbers ③	**20**	61	11	61
	30, 40, 60	66	7	69, 66, 70
	50, 70, 80	67	2	67, 72, 67
	90	68	9	68
teen numbers ④	**10**	46	5	47
	11 "eleven"	43	2 and 5	45
	12 "twelve"	43	2 and 5	45
	13 "thir-"	41	2	41
	14	35	2 and 8	37
	15 "fif-"	38	1 and 5	38
	16–19	35	2 and 8	36
one hundred numbers	**100**	76	5	77
	101–120	82	2	82
	121–190s	79	2	81
other symbols	= "equals"	5	2	7
	——— "equals bar"	85	4	(not written)
	+ "plus"	15	2	19
	− "take away"	35	6	47
	□ "box"	6	2	(not written)
	□ "how many"	59	4	(not written)
counters	I (ones counter)	15	5 and 9	17
	T (tens counter)	60	6	63

As the table shows, the schedule for identifying numbers does not follow the counting order, nor does it parallel the schedule for when the symbols are written (however, introduction of a symbol always precedes writing of the symbol).

IDENTIFYING SINGLE-DIGIT SYMBOLS

The first symbol introduced is 4. Here's the exercise from Lesson 1:

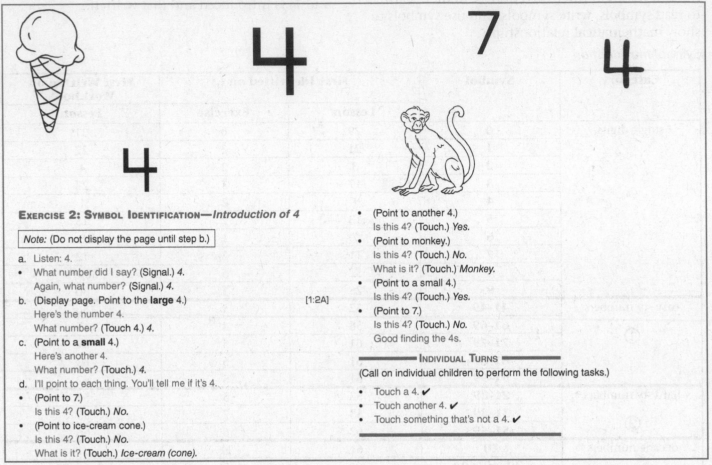

EXERCISE 2: SYMBOL IDENTIFICATION—*Introduction of 4*

> *Note:* (Do not display the page until step b.)

a. Listen: 4.
  - What number did I say? (Signal.) *4.*
    Again, what number? (Signal.) *4.*
b. (Display page. Point to the **large** 4.)   [1:2A]
   Here's the number 4.
   What number? (Touch 4.) *4.*
c. (Point to a **small** 4.)
   Here's another 4.
   What number? (Touch.) *4.*
d. I'll point to each thing. You'll tell me if it's 4.
  - (Point to 7.)
    Is this 4? (Touch.) *No.*
  - (Point to ice-cream cone.)
    Is this 4? (Touch.) *No.*
    What is it? (Touch.) *Ice-cream (cone).*

- (Point to another 4.)
  Is this 4? (Touch.) *Yes.*
- (Point to monkey.)
  Is this 4? (Touch.) *No.*
  What is it? (Touch.) *Monkey.*
- (Point to a small 4.)
  Is this 4? (Touch.) *Yes.*
- (Point to 7.)
  Is this 4? (Touch.) *No.*
  Good finding the 4s.

——————— **INDIVIDUAL TURNS** ———————
(Call on individual children to perform the following tasks.)

- Touch a 4. ✔
- Touch another 4. ✔
- Touch something that's not a 4. ✔

Lesson 1, Exercise 2

In step A, the teacher models saying 4, and the children say 4. In step B, the teacher displays the page, identifies an example of 4, and asks the children, "What number?" In step C, the teacher repeats step B with another 4. In step D, the teacher points to the objects on the page and asks, "Is this 4?" After children are firm on step D, the teacher presents Individual Turns.

Teaching Note: When presenting the tasks in steps B through D, make sure you first point to the symbol, then identify it or ask the question. Don't talk about an object before you point to it.

After asking the question, touch the symbol to signal when children are to respond. Use the same timing for touching the symbol that you use for signaling children to say the next number.

Praise children for correct responses. If children make a mistake, promptly tell them the answer and then repeat the task they missed.

Try to present the Individual Turns quickly. If possible, hold the book close to the child who is to respond, so the child does not have to walk to the book. Or use the online Board Displays.

Connecting Math Concepts

Later in the lesson, another exercise presents 4s and things that are not 4. Children practice identifying 4 on Lesson 2 and on Lesson 3. On Lesson 3 the next symbol, 2, is introduced and children identify both 4 and 2.

Variations of the procedure used to introduce 4 are used to introduce later symbols. As children learn more symbols, the symbols that children discriminate become increasingly similar; for instance, **8** is not introduced until children have been identifying **3** for 15 lessons.

Here's the introduction of 8 on Lesson 32:

EXERCISE 5: SYMBOL IDENTIFICATION—*Introduction of 8*

a. (Display page and point to ⟨8⟩.)   [32:5A]
   Here's a new number. This is 8.
   What is it? (Touch.) *8.*
• (Point to 8.) Is this 8? (Touch.) *Yes.*
   What is it? (Touch.) *8.*

b. (Point to **3**.) Is this 8? (Touch.) *No.*
   What is it? (Signal.) *3.*
c. (Repeat step b for the following:)

(Point to ___.)	Is this 8?	What is it?
6	*No*	*6*
0	*No*	*Zero*
8	*Yes*	*8*
9	*No*	*9*

Lesson 32, Exercise 5

All examples that are not 8 are symbols children have learned and practiced identifying for at least three lessons.

In step A, the teacher identifies the 8 in the box and asks, "What is it?" For the rest of step A and step B, the teacher points to each symbol and asks two questions:

• Is this 8?
• What is it?

In step C, the teacher points to the symbols in the reverse order and children identify each symbol.

Teaching Note: For steps B and C, point to the symbol before asking a question about the symbol. When you touch the symbols after the question, children are to respond. Use the same timing you've established in counting tasks.

If children make a mistake in step C, correct it immediately. Then provide children with additional time to think for each symbol. Point to a symbol and pause longer before asking, "What is this?"

Later children will write single digits. Reversed digits such as Ͱ for 4 are acceptable.

TWO-PART NUMBERS

> Reversed digits (⊽ for 7) are acceptable. However, transposed digits (72 for 27) are not acceptable. **Do not accept transposed digits for two-digit numbers.**

The word *digits* is not introduced in *CMC Level A*. Two-digit numbers are *two-part numbers*. The tens digit is referred to as *the first part* and the ones digit is referred to as *the other part*. Instruction on two-digit numbers does not begin until children have learned and practiced identifying all one-digit numbers for at least three lessons. Throughout *CMC Level A*, children are never expected to identify or write numbers that haven't been taught first. Before children learn to read numbers like 47, they receive a great deal of practice reading the single-digit numbers that compose 47: 4 and 7. Before children are expected to read or write 47 in a problem-solving or computational task, children will have practiced reading and writing 47. To make sure children don't write incorrect numbers, the teacher often prompts children—by writing the number on the board after children say what they are to write.

Two-digit numbers are grouped for instruction in four separate categories:

Category 1: Easy -ty numbers. One instructional category contains the numbers in the 40s, 60s, 70s, 80s, and 90s that do not end in zero. These numbers have a very high correspondence between the name and digits. The name for 64 contains the number 6 and 4 separated by a syllable, *ty*—sixty-four. This correspondence between the name and the digits works for all numbers in category 1.

Category 2: Hard -ty numbers. Another category is two-digit numbers that do not end in zero for 20s, 30s, and 50s. The correspondence between the name and the digits is high with predictable relationships. The name of the number tells about the two digits but the name for the tens digit is distorted. For all 50s numbers, the name *five* is replaced with *fif*. For the number *57*, the name is *fifty-seven*. For all 30s numbers, the name *three* is replaced with *thir*. For all 20s numbers, the name *two* is replaced with *twen*.

Category 3: Decade numbers. The decade numbers, 20, 30, 40, 50, 60, 70, 80, and 90, fall into another instructional category. These numbers do not follow the same pattern as two-digit numbers that do not end in zero. The names for decade numbers only contain the name that identifies the first digit. If the number *40* followed the same pattern as instructional categories 1 and 2, the name would be *forty-zero*, not just *forty*.

Category 4: Teen numbers. The last instructional category is the teen numbers, 10–19. The names for teen numbers have poor correspondence to the digits in the numbers. The number 14, first name contains the second digit, *four*, and contains a reference to the first digit, *teen,* but the parts are in the wrong order. The reference, *teen*, is not similar to the name of the first digit, *one*. The names for the teen numbers 13 and 15 have even less correspondence to the digits than the names for 14, 16, 17, 18, and 19. *Thir* and *fif* replace *three* and *five* respectively. The names for 11 and 12 have almost no correspondence to the digits.

The teen numbers that have names with a higher correspondence to the digits (14, 16, 17, 18, 19) are first taught on Lesson 35.

Here's the introduction to teen numbers from
Lesson 35:

EXERCISE 2: TWO-DIGIT NUMBERS—*Teens*

a. (Write on the board:)   [35:2A]

| | 7 | 4 | 9 | 6 | |

Read these numbers.
- (Point to **7.**) Get ready. (Touch.) *7.*
- (Point to **4.**) Get ready. (Touch.) *4.*
- (Point to **9.**) Get ready. (Touch.) *9.*
- (Point to **6.**) Get ready. (Touch.) *6.*

b. I'm going to change these numbers into teen numbers.
- (Point to **7.**) What's this number? (Signal.) *7.*
  (Write to show:)   [35:2B]

| | 17 | 4 | 9 | 6 | |

- Now it's not 7; it's 17.
  What is it? (Signal.) *17.*
- (Erase **1** to show:)   [35:2C]

| | 7 | 4 | 9 | 6 | |

- What is it now? (Signal.) *7.*
  (Write **1** to show:)   [35:2D]

| | 17 | 4 | 9 | 6 | |

- What is it now? (Signal.) *17.*

c. (Point to **4.**) What number? (Signal.) *4.*
  (Write to show:)   [35:2E]

| | 17 | 14 | 9 | 6 | |

- What number is it now? (Signal.) *14.*

d. (Point to **9.**) What is this number? (Signal.) *9.*
  (Write to show:)   [35:2F]

| | 17 | 14 | 19 | 6 | |

- What number is it now? (Signal.) *19.*

e. (Point to **6.**) What is this number? (Signal.) *6.*
  (Write to show:)   [35:2G]

| | 17 | 14 | 19 | 16 | |

- What number is it now? (Signal.) *16.*

f. My turn to read these numbers: (Touch each number.) *17, 14, 19, 16.*
  Your turn to read these numbers.
- (Point to **17.**) Get ready. (Touch.) *17.*
- (Point to **14.**) Get ready. (Touch.) *14.*
- (Point to **19.**) Get ready. (Touch.) *19.*
- (Point to **16.**) Get ready. (Touch.) *16.*
  (Repeat step f until firm.)

Lesson 35, Exercise 2

In step A, children identify single-digit numbers that they already know. In step B, children identify 7 again, and the teacher changes it to 17 and identifies it. Then, children discriminate between 17 and 7 as the teacher erases and rewrites 1. In steps C, D, and E, children identify single-digit numbers. Then the teacher writes a 1 in front of them, and children identify the teen numbers. In step F, the teacher reads all the teen numbers. Then children read the teen numbers.

On Lesson 36, children identify one-digit numbers in their Workbook and change them into teen numbers.

Here's the Workbook part from Lesson 36:

____ 6 ____ 7 ____ 8 ____ 9

Lesson 36, Workbook

On Lesson 38, the first teen number with a somewhat distorted name is introduced—15.

On Lesson 40, children learn to discriminate where to write 1.

Here's the exercise from Lesson 40:

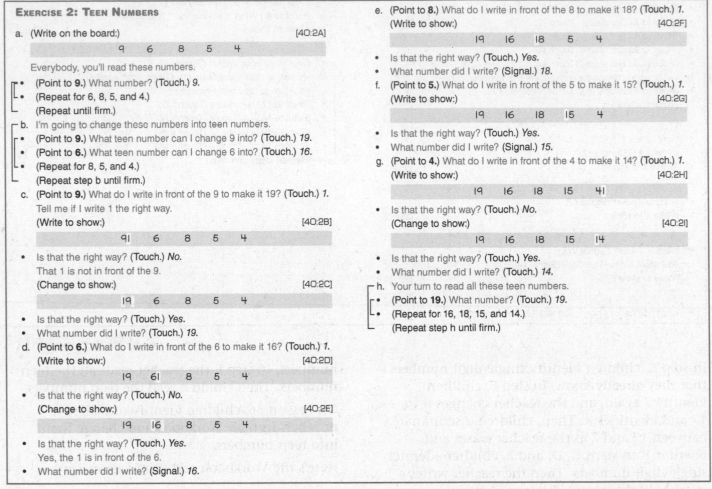

Lesson 40, Exercise 2

In step A, children identify five single-digit numbers. In step B, children identify the teen number that each single-digit number can be transformed into. In step C, children answer the question "What do I write in front of the 9 to make it 19?" Then the teacher writes 1 after 9 and asks, "Is that the right way?" The teacher erases the 1 after 9 and writes 1 in front of 9 and asks, "Is that the right way?" In steps C through G, the teacher presents similar tasks and discriminations for the remaining single-digit numbers. In step H, children read the teen numbers.

Teaching Note: Children are expected to answer only the questions that are indicated in the exercise. They are not expected to identify the numbers that are NOT written the right way. The names for these numbers will be introduced systematically later in the program. If children can identify whether the 1 is written the right way or not, they have a concrete test for identifying and writing teen numbers.

Children identify 13 on Lesson 41. On Lesson 43, 11 and 12 are introduced on a number line with 13, 14, and 15. On Lesson 46, the number 10 is introduced. Children practice identifying teen numbers and discriminating teen numbers from numbers that are not teens in every lesson through Lesson 59.

Here's an exercise from Lesson 52 that firms the test children learn for discriminating teen numbers:

31   61   17   12

41   13   10   **81**

EXERCISE 3: TWO-DIGIT NUMBERS—*Teen Discrimination*

a. (Display page and point to numbers.) [52:3A]
All of these numbers have two parts. One part of each number is 1, but not all these numbers are teen numbers. Remember, the first part of a teen number is 1.

• (Point to 31.) What's the first part of this number? (Touch.) *3.*
Is this a teen number? (Touch.) *No.*
• (Point to 61.) What's the first part of this number? (Touch.) *6.*
Is this a teen number? (Touch.) *No.*
• (Point to 17.) What's the first part of this number? (Touch.) *1.*
Is this a teen number? (Touch.) *Yes.*
What number is it? (Touch.) *17.*
• (Point to 12.) What's the first part of this number? (Touch.) *1.*
Is this a teen number? (Touch.) *Yes.*
What number is it? (Touch.) *12.*

• (Point to 41.) What's the first part of this number? (Touch.) *4.*
Is this a teen number? (Touch.) *No.*
• (Point to 13.) What's the first part of this number? (Touch.) *1.*
Is this a teen number? (Touch.) *Yes.*
What number is it? (Touch.) *13.*
• (Point to 10.) What's the first part of this number? (Touch.) *1.*
Is this a teen number? (Touch.) *Yes.*
What number is it? (Touch.) *10.*
• (Point to 81.) What's the first part of this number? (Touch.) *8.*
Is this a teen number? (Touch.) *No.*
(Repeat step a until firm.)

Lesson 52, Exercise 3

For each two-digit number, the teacher asks two questions:

• What's the first part of this number?
• Is this a teen number?

For the teen numbers, the teacher asks a third question:

• What number is it?

Teaching Note: Children shouldn't have any trouble with the first question. If children make mistakes on the second question, tell them, "This is a teen number" or "This is not a teen number." Then repeat the first question. Before asking the second question, ask, "Is the first part 1?" Then ask the second question. The questions are parallel, so if they answer "Yes" to the inserted question, they answer "Yes" to the second question.

If children make mistakes on the third question, tell them the answer and repeat the task.

After children finish the exercise, repeat the tasks for the numbers on which children made mistakes.

Later, on Lesson 52, two-digit numbers that are not teen numbers are introduced:

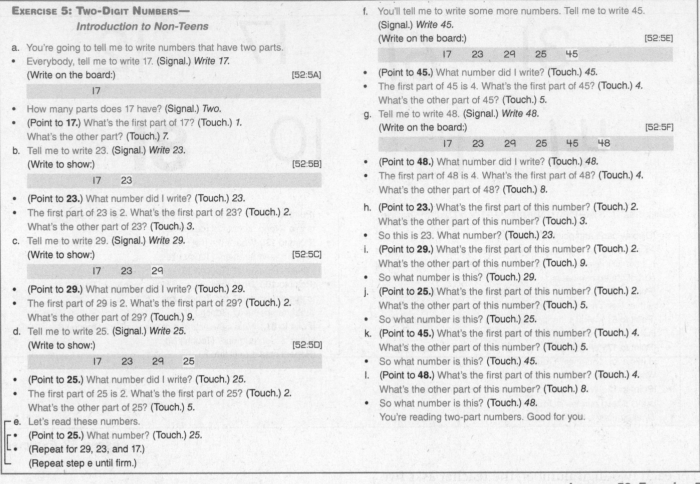

EXERCISE 5: TWO-DIGIT NUMBERS—
Introduction to Non-Teens

a. You're going to tell me to write numbers that have two parts.
• Everybody, tell me to write 17. (Signal.) *Write 17.*
(Write on the board:)   [52:5A]

| 17 |

• How many parts does 17 have? (Signal.) *Two.*
• (Point to **17**.) What's the first part of 17? (Touch.) *1.*
What's the other part? (Touch.) *7.*
b. Tell me to write 23. (Signal.) *Write 23.*
(Write to show:)   [52:5B]

| 17 | 23 |

• (Point to **23**.) What number did I write? (Touch.) *23.*
• The first part of 23 is 2. What's the first part of 23? (Touch.) *2.*
What's the other part of 23? (Touch.) *3.*
c. Tell me to write 29. (Signal.) *Write 29.*
(Write to show:)   [52:5C]

| 17 | 23 | 29 |

• (Point to **29**.) What number did I write? (Touch.) *29.*
• The first part of 29 is 2. What's the first part of 29? (Touch.) *2.*
What's the other part of 29? (Touch.) *9.*
d. Tell me to write 25. (Signal.) *Write 25.*
(Write to show:)   [52:5D]

| 17 | 23 | 29 | 25 |

• (Point to **25**.) What number did I write? (Touch.) *25.*
• The first part of 25 is 2. What's the first part of 25? (Touch.) *2.*
What's the other part of 25? (Touch.) *5.*
e. Let's read these numbers.
• (Point to **25**.) What number? (Touch.) *25.*
• (Repeat for 29, 23, and 17.)
(Repeat step e until firm.)

f. You'll tell me to write some more numbers. Tell me to write 45.
(Signal.) *Write 45.*
(Write on the board:)   [52:5E]

| 17 | 23 | 29 | 25 | 45 |

• (Point to **45**.) What number did I write? (Touch.) *45.*
• The first part of 45 is 4. What's the first part of 45? (Touch.) *4.*
What's the other part of 45? (Touch.) *5.*
g. Tell me to write 48. (Signal.) *Write 48.*
(Write on the board:)   [52:5F]

| 17 | 23 | 29 | 25 | 45 | 48 |

• (Point to **48**.) What number did I write? (Touch.) *48.*
• The first part of 48 is 4. What's the first part of 48? (Touch.) *4.*
What's the other part of 48? (Touch.) *8.*
h. (Point to **23**.) What's the first part of this number? (Touch.) *2.*
What's the other part of this number? (Touch.) *3.*
• So this is 23. What number? (Touch.) *23.*
i. (Point to **29**.) What's the first part of this number? (Touch.) *2.*
What's the other part of this number? (Touch.) *9.*
• So what number is this? (Touch.) *29.*
j. (Point to **25**.) What's the first part of this number? (Touch.) *2.*
What's the other part of this number? (Touch.) *5.*
• So what number is this? (Touch.) *25.*
k. (Point to **45**.) What's the first part of this number? (Touch.) *4.*
What's the other part of this number? (Touch.) *5.*
• So what number is this? (Touch.) *45.*
l. (Point to **48**.) What's the first part of this number? (Touch.) *4.*
What's the other part of this number? (Touch.) *8.*
• So what number is this? (Touch.) *48.*
You're reading two-part numbers. Good for you.

Lesson 52, Exercise 5

In step A, the teacher says "Everybody, tell me to write it." The teacher asks what number she wrote, what the first part of the number is, and what the other part is. The numbers the teacher writes for steps B through D are 20s numbers that are new: 23, 29, 25. In step E, children read the four numbers. Steps F and G are the same steps as A through D for new numbers 45 and 48. In steps H through L, children answer the following questions about the new numbers:

• What's the first part of this number?
• What's the other part of this number?
• So what number is this?

Teaching Note: If children are firm on identifying single-digit numbers, they should have little trouble with this exercise. Even though the digits for 20s numbers don't correspond identically to the parts in their names, the children are directed to say each digit before the teacher writes it.

20s, 30s, 40s, and 50s numbers are introduced and reviewed in similar exercises for the next 20 lessons. Simultaneously, children learn the digits for numbers that have a high correspondence between the digits and the name.

Here's the introduction from Lesson 58:

EXERCISE 4: SAYING SYMBOLS FOR TWO-DIGIT NUMBERS

a. You can figure out the symbols for two-part numbers that you've never read before.
- Listen: 68. What number? (Signal.) *68.*
  My turn: What's the first part of **68**? 6.
  What's the other part of **68**? 8.
- Your turn: What's the first part of **68**? (Signal.) *6.*
  What's the other part of **68**? (Signal.) *8.*
  (Repeat until firm.)
b. Listen: 64. What number? (Signal.) *64.*
- What's the first part of **64**? (Signal.) *6.*
  What's the other part of **64**? (Signal.) *4.*
c. Listen: 84. What number? (Signal.) *84.*
- What's the first part of **84**? (Signal.) *8.*
  What's the other part of **84**? (Signal.) *4.*
d. (Repeat the following tasks for 81, 73, 49:)
- Listen: ____. What number?
- What's the first part of ____?
- What's the other part of ____?

Lesson 58, Exercise 4

In step A, the teacher models the answer to the questions:

- What's the first part of **68**? 6.
- What's the other part of **68**? 8.

Then children respond to those questions. In steps B through D, the teacher says a number (64, 84, 81, 73, 49) and asks three questions:

- What number?
- What's the first part of (the number)?
- What's the other part of (the number)?

Teaching Note: This task is oral and doesn't require children to identify symbols, but since they know how to write the symbols that are named, they are able to use the name as a map for writing the number.

The answer to every question that asks about a part is imbedded in the question itself. If the teacher appropriately emphasizes the part of the question that is the answer, it is much easier for children to respond correctly. It's easier for children to answer 6 if the 6 in **six**ty-four is exaggerated when you ask the question, "What's the first part of **64**?"

Caution: If parts are overexaggerated, it can be difficult for children to recognize the number the teacher is asking about. Overexaggeration can also disrupt the timing of the questions.

Practice this exercise, emphasizing the parts while keeping the numbers recognizable and the tasks flowing.

Children must be firm on the oral skill of identifying the parts of regular two-digit numbers before they become fluent writing these numbers. Firm these oral exercises because it will save time later.

Children struggle writing two-digit numbers for two main reasons: They have trouble identifying the parts to write; they have trouble writing the parts. If struggling children are not reliable in identifying the parts, the teacher will not know what error children are making—identifying the parts or writing them. If children are firm on identifying the parts but are having trouble writing two-digit numbers, the teacher can conclude children need practice writing the parts.

Teen numbers, hard -ty numbers, (**twen**ties, **thir**ties, **fif**ties) and regular (easy) two-digit numbers (**for**ties, **six**ties, **seven**ties, **eigh**ties, **nine**ties) are systematically combined. Here's an exercise from Lesson 64 that combines teen numbers with 20s, 30s, and 50s numbers:

EXERCISE 2: SYMBOLS FOR TWO-DIGIT NUMBERS—
Saying and Writing

a. You'll tell me the symbols for teen numbers, 20s numbers, and 30s numbers, and I'll write them.
- Everybody, tell me the first part of teen numbers. Get ready. (Signal.) *1.*
- Tell me the first part of 50s numbers. Get ready. (Signal.) *5.*
- Tell me the first part of 20s numbers. Get ready. (Signal.) *2.*
- Tell me the first part of 30s numbers. Get ready. (Signal.) *3.*
  (Repeat until firm.)
b. Listen: 11. What number? (Signal.) *11.*
- What's the first part of 11? (Signal.) *1.*
  What's the other part of 11? (Signal.) *1.*
- Say both parts of 11. (Signal.) *1 and 1.*
  (Write on the board:)   [64:2A]

  | 11 |

- (Point to **11**.) What number? (Touch.) *11.*
c. Next number: 52. What number? (Signal.) *52.*
- What's the first part of 52? (Signal.) *5.*
  What's the other part of 52? (Signal.) *2.*
- Say both parts of 52. (Signal.) *5 and 2.*
  (Write to show:)   [64:2B]

  | 11 | 52 |

- (Point to **52**.) What number? (Touch.) *52.*
d. Next number: 25. What number? (Signal.) *25.*
- What's the first part of 25? (Signal.) *2.*
  What's the other part of 25? (Signal.) *5.*
- Say both parts of 25. (Signal.) *2 and 5.*
  (Write to show:)   [64:2C]

  | 11 | 52 | 25 |

- (Point to **25**.) What number? (Touch.) *25.*

Lesson 64, Exercise 2

e. Next number: 34. What number? (Signal.) *34.*
• What's the first part of 34? (Signal.) *3.*
  What's the other part of 34? (Signal.) *4.*
• Say both parts of 34. (Signal.) *3 and 4.*
  (Write to show:)   [64:2D]

11	52	25	34

• (Point to 34.) What number? (Touch.) *34.*
f. You're going to read these numbers.
• (Point to **11**.) What number is this? (Touch.) *11.*
• (Point to **52**.) What number is this? (Touch.) *52.*
• (Point to **25**.) What number is this? (Touch.) *25.*
• (Point to **34**.) What number is this? (Touch.) *34.*

Lesson 64, Exercise 2 (continued)

In step A, children identify the first part of the numbers in categories 2 and 4: 20s, 30s, and 50s numbers and teen numbers. In steps B through E, the teacher says a number (11, 52, 25, 34) and the children respond to four tasks:

• What number?
• What's the first part of (the number)?
• What's the other part of (the number)?
• Say both parts of (the number).

In step F, children read the numbers.

Teaching Note: Step A is very important. In step A, children identify the first part for teen numbers, 20s numbers, 30s numbers, and 50s numbers. For each task in step A, pause long enough before saying "Get ready" to allow children who know the answer to respond together. Use the timing you've established for all other tasks after you say "Get ready."

In step C, read **52** as **fif**ty-two and **52** as fifty-**two**. In step D, read **25** as **twenty**-five, and in step E read **34** as **thir**ty-four.

Directing children to say both parts of numbers is new. Gradually, the questions about the parts will be faded out, and children will be asked to say both parts for numbers. Children will apply their number skills extensively when they write two-digit numbers as part of problem-solving and computational operations.

Decade numbers (10, 20, 30, 40, 50, 60, 70, 80, 90) are introduced on Lesson 66. The following exercise from Lesson 67 shows the relationship between writing familiar two-digit numbers and the decade number that starts with the same first digit.

Here's the exercise from Lesson 67:

EXERCISE 2: SYMBOL IDENTIFICATION—*Tens Numbers*

a. You'll tell me the symbols for two-part numbers, and I'll write them. Then you'll tell me what to write, and I'll write the tens numbers below.
• Listen: 78. What number? (Signal.) *78.*
• Think of the first part of 78 and the other part of 78. What are parts for 78? (Signal.) *7 and 8.*
  (Write on the board:)   [67:2A]

78

• (Point to **78**.) Below 78, I'm going to write 70. What number am I going to write below 78? (Touch.) *70.*
• Everybody, what's the first part of **seven**ty? (Signal.) *7.*
  (Write to show:)   [67:2B]

78
7

• The other part of 70 is zero. What's the other part of 70? (Signal.) *Zero.*
  (Write to show:)   [67:2C]

78
70

• (Point to **70**.) What number? (Touch.) *70.*
• (Point to **78**.) What number? (Touch.) *78.*
b. New number: 51. What number? (Signal.) *51.*
• Think of the first part of 51 and the other part of 51. What are parts for 51? (Signal.) *5 and 1.*
  (Write to show:)   [67:2D]

78	51
70	

• (Point to **51**.) Below 51, I'm going to write 50. What number am I going to write below 51? (Touch.) *50.*
• Everybody, what's the first part of 50? (Signal.) *5.*
  (Write to show:)   [67:2E]

78	51
70	5

• What's the other part of 50? (Signal.) *Zero.*
  (Write to show:)   [67:2F]

78	51
70	50

• (Point to **50**.) What number? (Signal.) *50.*
• (Point to **51**.) What number? (Touch.) *51.*
c. I'll say the parts for one of the numbers on the board. You'll say those parts, then tell me the number.
• The parts for one of the numbers are 5 and zero. Say those parts. (Signal.) *5 and zero.*
• Everybody, what number? (Signal.) *50.*
  (Touch **50**.) Yes, the parts for 50 are 5 and zero.
d. The parts for one of the numbers are 7 and zero. Say those parts. (Signal.) *7 and zero.*
• Everybody, what number? (Signal.) *70.*
  (Touch **70**.) Yes, the parts for 70 are 7 and zero.
e. The parts for one of the numbers are 5 and 1. Say those parts. (Signal.) *5 and 1.*
• Everybody, what number? (Signal.) *51.*
  (Touch **51**.) Yes, the parts for 51 are 5 and 1.

Lesson 67, Exercise 2

Connecting Math Concepts

Similar exercises for teaching children how to read and write decade numbers continue for several lessons. Work on all four categories of two-digit numbers continues until children are practiced enough to discriminate numbers in each category and use numbers from 1–99 in the problem-solving and computational operations taught in *CMC Level A*.

THREE-PART NUMBERS

Work with numbers 101–199 begins on Lesson 76. Children learn to identify 100 in Lesson 76 and write it in Lesson 77. The "easy" numbers, 121–190s, are introduced first (Lesson 79). The "hard" numbers, 101–120, are introduced beginning on Lesson 82.

> **Teaching Note:** Read **100** as "one hundred," not "**a** hundred," and read **121** as "one hundred 21," not "one hundred **and** 21."

OTHER SYMBOLS (=, +, –, □)

CMC Level A introduces symbols that are not numbers: =, +, –, and □.

These symbols occur in the context of operations. Equals occurs in a variety of operations starting with problems as simple as ||||=. Children count the lines, and write the number 4 on the other side to make the sides equal.

Children work with the symbols +, –, and □ in problem-solving and computational operations. Children must be able to reliably identify the symbols and understand what they mean.

Children first learn to identify the symbols, and separately learn what the symbols mean before the symbols appear in the context of operations.

Here's the last part of the exercise from Lesson 15, which introduces the + symbol:

$$+ \; 4 \; 5 \; = \; 7 \; + \; 5$$

b. (Display page and point to +.)   [15:2B]
This is **plus.** What is it? (Touch.) *Plus.*
Yes, plus.

from Lesson 15, Exercise 2

> **Teaching Note:** You and the children say *plus* for +. You don't say **and**. Likewise, – is taught as *take away*, not as **minus**. (In *CMC Level B* the term *minus* is introduced.)

Equality and Equations— Lessons 19–120

The equals symbol (=) is introduced on Lesson 5. "=" is not used in operations until Lesson 19. First, children learn and apply the basic rule about equality. **You have the same number on one side of the equals that you have on the other side.** Children first apply this rule to problems of the form 4 = □ and □ = 7. Then children work with problems that have counters on one side of the equals sign and a number on the other side. The different types of problems children work include:

- Making lines to complete equations that have a number shown: 4 =
- Writing numbers to complete equations that have lines shown: |||||=
- Identifying whether equations are correct or not: 4 = ||
- Writing numbers that have lines shown, but some of the lines are crossed out: ||||H =
- Writing numbers to complete equations that have Ts and lines shown: TTTT|||=
- Writing numbers that have Ts and lines shown, but some of the Ts or lines are crossed out: TTTT|||=

After children master working each of these problem types, they apply those skills to higher-order problems. Over the course of the *CMC Level A* program, children's understanding of these basic relationships of equality is expanded to include all of the facts, computational operations, and problem-solving types that *CMC Level A* presents.

THE RULE ABOUT EQUALITY

On Lesson 19, the operational rule for equality is introduced. Here's the first part of the exercise from Lesson 19:

EXERCISE 6: EQUALS

a. (Write on the board:)   [19:6A]

- (Point to =.) Everybody, what's this? (Touch.) *Equals.*
- Is equals a number? (Touch.) *No.*
  Right. It isn't.
- Here's the rule about equals. (Point to the space on the left.) You must have the same number on this side of the equals (point to the right side of the equals) and on this side of the equals.

b. Listen. (Point left.) If we have 4 on this side of the equals, (point right) we must have 4 on this side of the equals.
- (Point left.) Listen: If we have 4 on this side of the equals, (point right) how many must we have on this side of the equals? (Touch.) *4.*
  Yes, 4.

c. (Point right.) If we have 10 on this side of the equals, (point left) how many must we have on this side of the equals? (Touch.) *10.*
- (Point left.) If we have 15 on this side of the equals, (point right) how many must we have on this side of the equals? (Touch.) *15.*
  (Repeat steps b and c until firm.)

d. (Write to show:)   [19:6B]

$$4 = \square$$

- This says **4 equals box.** What does it say? (Touch.) *4 equals box.*
- There's a number on one side of the equals. What number? (Signal.) *4.*
  Yes, 4.

e. (Point to **4.**) If we have 4 on that side of the equals, (point to box) how many must there be on this side of the equals? (Touch.) *4.*
  Yes, that's the number that goes in the box.
- (Write to show:)   [19:6C]

$$4 = \boxed{4}$$

(Point to 4 = 4.) Now this says (touch each symbol as you read) 4 equals 4.
- What does it say? (Touch symbols.) *4 equals 4.*

f. (Write on the board:)   [19:6D]

$$\square = 2$$

This says **box equals 2.**
- There's a number on one side of the equals. What number? (Signal.) *2.*
- (Point to 2.) If there are 2 on this side of the equals, (point to box) how many must there be on this side of the equals? (Touch.) *2.*
  Yes, that's the number that goes in the box.
- (Write to show:)   [19:6E]

$$\boxed{2} = 2$$

- (Point to 2 = 2.) What does this say now? (Touch symbols.) *2 equals 2.*

from Lesson 19, Exercise 6

In steps A through C, the teacher writes (or displays) an equals sign and points to one of the sides. The teacher asks the children, "If we have __ on this side of the equals, how many must we have on this side?" as she points to the other side of the equals. In step D, the teacher displays: 4 = □. Children say the statement "4 equals box" and identify the number on one side of the equals. In step E, children identify how many must be on the other side (4). The teacher completes the equation and reads it. Then children read the equation. Step F presents the same tasks that are in steps D and E with a new problem: □ = 2.

The examples show that you can operate on either side of the equation to make the sides equal. Also, if a problem has a box, you write a number in the box to make the sides equal. Both conventions will be incorporated in most of the work that children do in *CMC Level A*.

> **Teaching Note:** Timing is important for these exercises. For steps A through C, say your lines quickly so children can hear the simple relationship. The longer the pauses, the less likely it is that children will grasp the relationship.
>
> Present steps D through F using the strategy for presenting symbol-identification exercises: Point before you talk about it or ask about it. Don't talk about an object before you point to it. After asking a question, touch it with the same timing used in saying the next-number tasks or identifying-a-symbol tasks.

> Steps D though F are easy for children if:
>
> - Children are firm on answering the questions in steps A through C
> - The tasks are presented at a brisk pace
> - Your pointing and touching is coordinated with your words
>
> Practice this exercise before presenting it to ensure that you are able to present the tasks at a brisk pace and coordinate the presentation with your point-and-touch signaling.

INEQUALITY DISCRIMINATION

On later exercises, children do a lot of work with equations that have lines on one side and a number on the other side. In the initial exercises, children are directed to identify the number on one side and count the lines on the other side. On Lessons 23 and 24, these identification exercises are oral. On Lesson 25, children cross out the equals sign for statements that have the wrong number of lines.

Here's an exercise from Lesson 25:

EXERCISE 8: EQUALITY

a. (Write on the board:) [25:8A]

$$||||||| = 5$$

You're going to tell me if the sides are equal.

- (Point to lines.) I'll touch the lines. You'll count them. Get ready. (Touch lines.) *1, 2, 3, 4, 5, 6, 7.*
- How many lines? (Touch.) *7.*
- Look at the other side. ✔
- (Point to **5.**) What number? (Touch.) *5.*
  It says 7 equals 5.
- So are the sides equal? (Touch.) *No.*
- The sides are not equal. So I'll cross out the equals. Watch.
  (Write to show:) [25:8B]

$$|||||| \neq 5$$

I made only one line when I crossed it out. This says 7 does not equal 5. That's right.

b. Touch the duck on your worksheet. ✔
  (Teacher reference:)

  🦆$5 = ||||| \; ||||||| = 9 \; |||| = 4$

  There are equals in the row next to the duck. A number is on one side of each equals, and lines are on the other side. You'll cross out an equals if the sides are not equal.
- Touch the first equals. ✔
  You'll touch and count the lines.
- Finger over the first line. ✔
- Get ready. (Tap 6.) *1, 2, 3, 4, 5, 6.*
- How many lines? (Signal.) *6.*

- Touch the number on the other side. ✔
- What number? (Signal.) *5.*
  It says 5 equals 6.
- So are the sides equal? (Signal.) *No.*
  So you cross out that equals.
c. Put your pencil on the big ball and make one cross-out line.
  (Observe children and give feedback.)
  (Teacher reference:)

  🦆$5 \neq |||||$

d. Touch the next equals. ✔
  You'll touch and count the lines.
- Finger over the first line. ✔
- Get ready. (Tap 7.) *1, 2, 3, 4, 5, 6, 7.*
- How many lines? (Signal.) *7.*
- Touch the number on the other side. ✔
- What number? (Signal.) *9.*
  It says 7 equals 9.
- So are the sides equal? (Signal.) *No.*
- Make one cross-out line through the equals.
  (Observe children and give feedback.)
e. Touch the last equals. ✔
  You'll touch and count the lines.
- Finger over the first line. ✔
- Get ready. (Tap 4.) *1, 2, 3, 4.*
- How many lines? (Signal.) *4.*
- Touch the number on the other side. ✔
- What number? (Signal.) *4.*
- What does it say? (Signal.) *4 equals 4.*
- So are the sides equal? (Signal.) *Yes.*
  The sides are equal, so you don't cross out the equals.

Lesson 25, Exercise 8

To determine if the sides are equal, children count the lines on one side, identify the number on the other side, and say the statement. If the statement is correct (5 = 5), the equals is left alone. If the statement is not correct (7 = 5), a single line is made through the equals.

In step A, the teacher presents an example on the board which has 7 lines on one side and the number 5 on the other side. After children determine that the statement is not correct, the teacher makes an inequality sign to show that the sides are not equal. In the following steps, children work three examples in their Workbook. The first two examples are not equations, and the children change the equals to an inequality sign. The last example is an equation, and the children leave the equal sign.

Teaching Note: The equals sign changes to an inequality sign when it is crossed out. Children do not learn the name for the inequality sign, but children learn a solid conceptual model for understanding equality and inequality.

This exercise requires children to combine several skills they have learned. Children locate the equals in each problem, accurately count the lines on one side, identify the symbol on the other side, say the statement for each problem, and determine if the statement is correct. If children have mastered each of these skills in isolation, they will perform the tasks in this exercise easily.

If children have trouble with this exercise, they have not mastered all the skills listed above. Identify the skill and firm it in isolation. Children's fluency and accuracy will increase as they master isolated skills. Mastering this problem type is an essential pre-skill for working the more advanced problems involving equations.

COMPLETING EQUATIONS

After children can make specified numbers of lines and have worked with inequality discriminations, children complete equations to show a group of lines for a number.

Here's part of the exercise and the Workbook activity for Equals on Lesson 27:

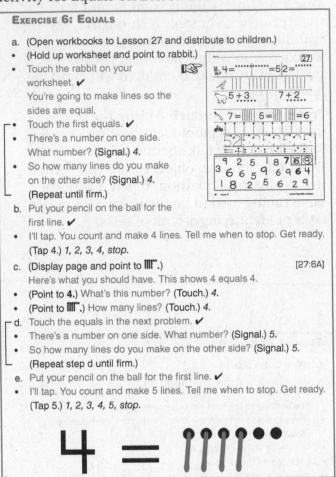

from Lesson 27, Exercise 6

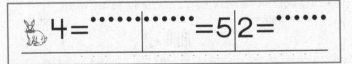

Lesson 27, Workbook

The number is shown on one side of equals. On the other side is a row of balls. Children use these as starting points for the lines they make, starting with the first ball. Children first identify the number and indicate how many lines they'll make on the other side. Then the teacher taps as the children count and make lines. After children make the last line, they tell the teacher to stop counting.

Before introducing a variation of the exercise above that presents lines on one side of the equals sign, children count the lines below boxes and write the number for the lines in each box. Children say the number names in the counting order, touching each line and pairing it with one and only one number name. Children identify the last number name they say as the number of lines in a group and write that number in the box. Children should be firm touching and counting objects. Children should also be firm identifying the last number name they say when they count the number of objects in the group. Writing the numbers in the boxes is the part of this procedure that is new to children.

Note that for the first exercises of this type, the teacher writes the symbol on the board after children identify the number of lines. The assistance children receive writing numbers decreases gradually and systematically.

Here's the first part of the exercise from Lesson 28 that introduces writing numbers for lines:

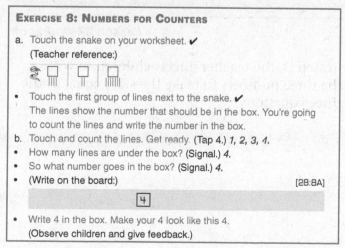

from Lesson 28, Exercise 8

In step B, children touch and count the lines, identify how many lines are under the box, and then identify the number that goes in the box. The teacher writes the number on the board and directs children to write the number.

Steps C and D (not shown here) present the same tasks for the remaining groups of lines under the boxes.

On Lesson 32, children work a mixed set of problems in which they write numbers for lines and make lines for numbers. Children apply the procedure they use to write numbers for lines when they complete equations.

Here's the Equations exercise from Lesson 37 and the Workbook part:

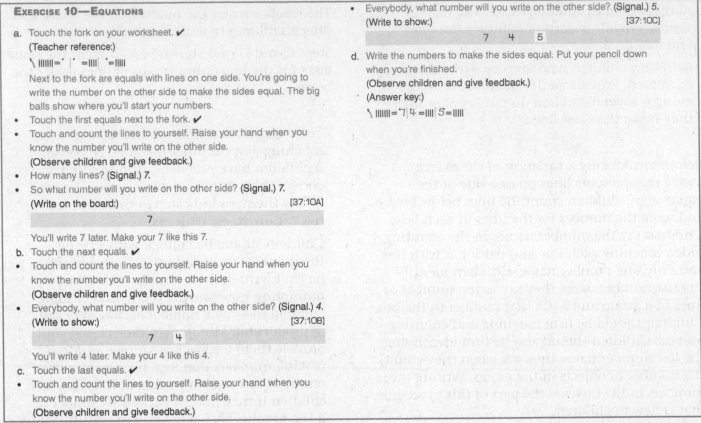

EXERCISE 10—EQUATIONS

a. Touch the fork on your worksheet. ✔
   (Teacher reference:)

   ＼ |||||| = ˙ | ˙ = |||| ˙ = ||||

   Next to the fork are equals with lines on one side. You're going to write the number on the other side to make the sides equal. The big balls show where you'll start your numbers.

   • Touch the first equals next to the fork. ✔
   • Touch and count the lines to yourself. Raise your hand when you know the number you'll write on the other side.
   (Observe children and give feedback.)
   • How many lines? (Signal.) 7.
   • So what number will you write on the other side? (Signal.) 7.
   (Write on the board:)   [37:10A]

   | 7 |

   You'll write 7 later. Make your 7 like this 7.

b. Touch the next equals. ✔
   • Touch and count the lines to yourself. Raise your hand when you know the number you'll write on the other side.
   (Observe children and give feedback.)
   • Everybody, what number will you write on the other side? (Signal.) 4.
   (Write to show:)   [37:10B]

   | 7   4 |

   You'll write 4 later. Make your 4 like this 4.

c. Touch the last equals. ✔
   • Touch and count the lines to yourself. Raise your hand when you know the number you'll write on the other side.
   (Observe children and give feedback.)

   • Everybody, what number will you write on the other side? (Signal.) 5.
   (Write to show:)   [37:10C]

   | 7   4   5 |

d. Write the numbers to make the sides equal. Put your pencil down when you're finished.
   (Observe children and give feedback.)
   (Answer key:)

   ＼ |||||| = ˙7 | 4 = |||| 5 = ||||

Lesson 37, Exercise 10

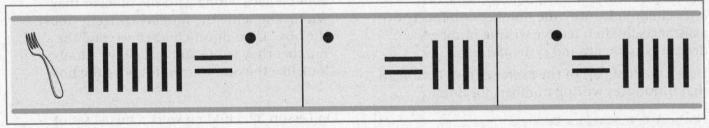

Lesson 37, Workbook

For each equation, children first count the lines to themselves, identify the number of lines and indicate what number they will write on the other side of the equals. After children identify the number, the teacher writes it on the board.

In step D, the teacher directs children to write the three numbers to make the sides equal in the three equations.

Teaching Note: For each example, children are to count the lines to themselves as you observe and give feedback. Some children may not be able to count lines to themselves if they don't vocalize the counting. Praise children who whisper or say the numbers softly.

When you observe and give feedback, don't interrupt children who are counting. After they are done, give them feedback. "Nice touching and counting to yourself." Also, ask them quietly, "How many lines are there?"

If children have problems counting the lines to themselves, direct the group to count the lines for each problem out loud. After children are firm counting the lines out loud, start the exercise again, and present it as specified.

Do not count reversed symbols (ᘐ) as wrong, but correct children by telling them that their number has to face the same way as the number you wrote. Tell them to cross out what they wrote and, above it, make the number that faces the same way your model faces. (See also Symbols, **Identifying Single-Digit Symbols,** page 66.)

Also on Lesson 37, children complete equations that have lines on one side, but some of the lines are crossed out. Children count the lines that are not crossed out to identify the number that makes the sides equal.

Beginning on Lesson 99, children work similar problems with Ts and lines. This work prepares children for using the same strategy for working two-digit subtraction problems that they use to work single-digit subtraction problems. Children count for the Ts that are not crossed out. Then they count for the lines.

Here are the problems that are presented in the Workbook on Lesson 99:

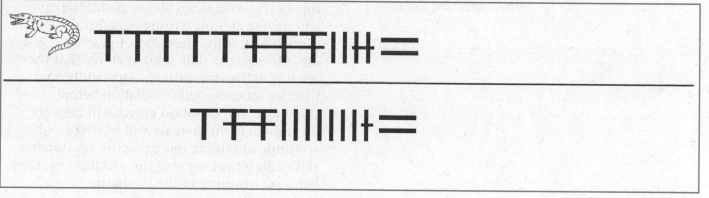

Lesson 99, Workbook

EQUATIONS FROM DICTATION

To assure that children learn the skills they will need later in the program to write word problems, the program first provides a simpler format for writing problems. Children write problems from dictation.

Here's part of an exercise from Lesson 36:

EXERCISE 8: PROBLEM DICTATION

a. Touch the flag on your worksheet. ✔
    (Teacher reference:)

    ✎ _____

    You'll write a problem that pluses. Then you'll work it.
    • Here's the problem: 3 plus 2. Say the problem. (Signal.) *3 plus 2.*
    • What symbol do you write first? (Signal.) *3.*
    • What do you write next? (Signal.) *Plus.*
    • What do you write next? (Signal.) *2.*
    (Repeat until firm.)
b. (Write on the board:)   [36:8A]

    3 + 2

    • Write 3 plus 2 on the line next to the flag. Make your 3 plus 2 look like this 3 plus 2.
    (Observe children and give feedback.)
c. We want to know what 3 plus 2 **equals.**
    • So what symbol do you write next? (Signal.) *Equals.*
    (Write to show:)   [36:8B]

    3 + 2 =

    • Write equals after 2.
    (Observe children and give feedback.)

---

from Lesson 36, Exercise 8

In step A, the teacher says the problem 3 + 2. Children say the problem, then answer questions about the symbol they write *first, next,* and *next.* In step B, the teacher writes the problem on the board and directs children to write it on their worksheet.

In steps C and D, the teacher models the next symbol, equals, and directs children to copy it. After the children write the problem, the teacher directs them to work it.

Teaching Note: By the time children write symbols for word problems (Lesson 47), they must be reasonably proficient at writing problems from dictation. Working word problems requires children to translate the word problem into a number problem, to dictate the number problem themselves, and then to write the problem.

This exercise focuses on several skills—repeating a problem that is presented orally, identifying the individual symbols in the problem, saying the symbols individually in the correct order, and writing the symbols.

However, if children have trouble with this exercise, they may need more work to master one of the skills listed above. If children are still having trouble writing equations from dictation after the third day of practice (Lesson 38), identify the skill(s) listed above that they are having trouble with. Provide additional practice on each skill in isolation before presenting the dictation exercise in Lessons 39 and 40. If children are still having trouble, continue providing more practice on isolated skills and repeat some of the dictation exercises before continuing in the program.

Place Value—Lessons 69–120

The work that children do with two-digit problems requires an understanding of place-value notations, namely that the first digit in a number like 36 tells the number of tens, and the other digit (6) tells the number of ones.

Before children begin working with place-value equations for two-digit numbers, they learn skills associated with place value in the Counting track and Symbols track. In the Counting track, they are taught that the number of Ts tells about the first digit and the number of lines tells about the other digit. (See pages 58–60, for example.) In the Symbols track, children learn that the names of most two-digit numbers tell about the place value. For example, the name for 36 tells that the value for the tens is 30 and the value for the ones is 6.

These values are confirmed by the representation of 36 in Ts and lines.

TTT||||||

Children count for the Ts: 10, 20, thirtyyy; then they count for the lines: 31, 32, 33, 34, 35, 36 and end up with the number represented by the numerical symbols. The skills learned in Counting and in Symbols prepare children for learning comprehensive place-value skills for all two-digit numbers.

In the Place-Value track, children learn to say the place-value equation for two-digit numbers. The place-value equation for 36 is 30 + 6 = 36. Children also do the actual addition and see that the equation is true. Later they work problems of the following type:

Children write the missing decade number and the missing ones number to complete the place-value equation.

For most two-digit numbers, the name is a clear guide for writing the place-value equations. For 86, you write 80 first, then 6. For 28, you write 20 first, then 8.

Teen and decade numbers are exceptions. For teens, the digits follow the reverse order of the name. For seventeen, you write 10 first, then 7. For 30 you write 30 first, then nothing.

The program does not introduce the place-value equations for these irregular numbers until children have practiced place-value equations that are regular for more than 30 lessons.

Here's the Numeral Expansion place-value exercise from Lesson 69 that introduces the verbal pattern for regular place-value equations:

EXERCISE 2: NUMERAL EXPANSION

a. We'll say tricky addition facts that start with tens numbers.
• Listen: 80 plus 7 equals 87.
• Say the fact for 80 plus 7. Get ready. (Signal.) *80 plus 7 equals 87.*

b. New fact: 70 plus 9 equals 79.
• Say the fact for 70 plus 9. Get ready. (Signal.) *70 plus 9 equals 79.* (Repeat step b until firm.)

c. This time I'll say problems. You tell me the answers.
• Listen: 20 plus 9. What does 20 plus 9 equal? (Signal.) *29.* Yes, 20 plus 9 equals 29.
• Say the fact for 20 plus 9. Get ready. (Signal.) *20 plus 9 equals 29.* (Repeat step c until firm)

d. Listen: 30 plus 8. What does 30 plus 8 equal? (Signal.) *38.* Yes, 30 plus 8 equals 38.
• Say the fact for 30 plus 8. Get ready. (Signal.) *30 plus 8 equals 38.*

e. Listen: 40 plus 1. What does 40 plus 1 equal? (Signal.) *41.* Yes, 40 plus 1 equals 41.
• Say the fact for 40 plus 1. Get ready. (Signal.) *40 plus 1 equals 41.* (Repeat steps d and e until firm.)

—— **INDIVIDUAL TURNS** ——
(Call on individual children to perform one or two of the following tasks.)
• Say the fact for 50 plus 7. (Call on a child.) *50 plus 7 equals 57.*
• Say the fact for 20 plus 6. (Call on a child.) *20 plus 6 equals 26.*
• Say the fact for 30 plus 2. (Call on a child.) *30 plus 2 equals 32.*

Lesson 69, Exercise 2

In steps A and B, the teacher models two facts; children say the facts: 80 + 7 = 87 and 70 + 9 = 79. In steps C through E, the teacher presents problems, and children say the answers. Then they say the facts. After step E, the teacher presents individual turns.

Teaching Note: For all of the initial place-value equations children say, just like with Counting, make the facts rhythmical, and make sure that children say them the same way you model facts in steps A and B. If children lose the beat, say to them after they say a fact:

"Yes, seventy plus **nine** equals seventy-**nine**."

"Say the fact for 70 plus 9 again."

After the group is firm on saying the place-value equations, call on children to verify individual mastery and give them the opportunity to show off their place-value skills.

After children have practiced saying place-value equations from problems for several lessons, they write answers to place-value problems in their Workbooks. In later lessons, children also generate place-value equations from 2-digit numbers. Example:

74. Say the place-value equation for 74. (Signal.) 70 + 4 = 74.

Here's the exercise from Lesson 79:

EXERCISE 9: NUMERAL EXPANSION—*Place-Value Preskill*

a. I'm going to say problems, and you'll tell me the whole equation.

- Listen: 80 plus 5. What's the answer to 80 plus 5? (Signal.) *85.* Yes, 80 plus 5 equals 85.
- Say the equation for 85. (Signal.) *80 plus 5 equals 85.*

b. New problem: 30 plus 7. What's the answer to 30 plus 7? (Signal.) *37.*

- Say the equation for 37. (Signal.) *30 plus 7 equals 37.*
- New problem: 20 plus 4. What's the answer to 20 plus 4? (Signal.) *24.*
- Say the equation for 24. (Signal.) *20 plus 4 equals 24.*
- New problem: 70 plus 1. What's the answer to 70 plus 1? (Signal.) *71.*
  Say the equation for 71. (Signal.) *70 plus 1 equals 71.*
  (Repeat steps a and b until firm.)

c. Now you're going to complete the equations for some of those numbers. Touch the number 71 on worksheet 79. ✔
(Teacher reference:)

$\square+\square=71$ | $\square+\square=24$

- Say the equation for 71 again. Get ready. (Tap 5.) *70 plus 1 equals 71.*
- Touch where you'll write 70. ✔
- Touch where you'll write 1. ✔
- Complete the equation for 71.
  (Observe children and give feedback.)

d. Touch the number 24. ✔

- Say the equation for 24 again. Get ready. (Tap 5.) *20 plus 4 equals 24.*
- Complete the equation for 24.
  (Observe children and give feedback.)
  (Answer key:)

$\boxed{70}+\boxed{1}=71$ | $\boxed{20}+\boxed{4}=24$

Lesson 79, Exercise 9

In steps A and B, the teacher says a place-value problem and asks, "What's the answer?" The teacher directs children to say the equation. In steps C and D, children go to their Workbook. The Workbook presents two of the place-value problems children worked in step B. For each problem, children say the place-value equation and complete it.

Later, children work sets of problems in their Workbook that contain both types of problems.

Here's the Workbook part from Lesson 81:

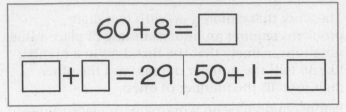

Lesson 81, Workbook

On Lesson 102, children begin work with place-value equations for teen numbers. On Lesson 110, children begin work with place-value equations for decade numbers. Before the end of the program, children complete place-value equations for sets that include teen numbers, decade numbers, and regular numbers.

Here's the Workbook part from Lesson 113: (The ones digit for decade problems is zero— 90 + 0 = 90.)

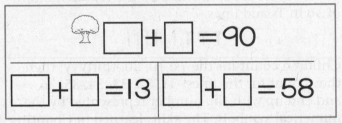

Lesson 113, Workbook

Plus (Addition)—Lessons 26–120

Children learn the properties of addition by counting groups. (See **Counting**, **Count On**, for a thorough discussion of counting groups.) Children count groups individually and count them together. Counting groups helps provide children with a conceptual understanding of the associative and commutative properties of addition. Children count groups of objects and obtain the same total number for the groups regardless of how the objects or groups are ordered.

After children have counted on for groups, and events, children translate addition problems into counting operations, and do the counting.

Here's the first part of the Plus exercise from Lesson 26:

EXERCISE 4: PLUS

a. I'm going to say problems that tell about two groups. You'll say each problem and tell me how many you plus.
- Listen: 2 plus 7. Say the problem. (Signal.) *2 plus 7.*
- How many does it tell you to plus? (Signal.) *7.*
- How do you plus 7? (Signal.) *Make 7 lines.*

b. New problem: 5 plus 1. Say the problem. (Signal.) *5 plus 1.*
- How many does it tell you to plus? (Signal.) *1.*
- How do you plus 1? (Signal.) *Make 1 line.*

c. New problem: 4 plus 3. Say the problem. (Signal.) *4 plus 3.*
- How many does it tell you to plus? (Signal.) *3.*
- How do you plus 3? (Signal.) *Make 3 lines.*

d. New problem: 8 plus 6. Say the problem. (Signal.) *8 plus 6.*
- How many does it tell you to plus? (Signal.) *6.*
- How do you plus 6? (Signal.) *Make 6 lines.*

e. (Write on the board:)   [26:4A]

> 2 + 3

- Read this part. Get ready. (Touch symbols.) *2 plus 3.*

f. (Point to **+ 3.**) This part tells me to plus some.
- How many does it tell me to plus? (Touch.) *3.*
- How do I plus 3? (Signal.) *Make 3 lines.*
  Yes, I make 3 lines under the 3.

g. I'll make lines. Count and tell me when to stop. Get ready. (Make 3 lines.) *1, 2, 3, stop.*
  (Teacher reference:)   [26:4B1–3]

> 2 + 3
> |||

h. Now you're going to count both groups.
- (Point to **2.**) How many are in this group? (Touch.) *2.*
- My turn to get 2 going and count the lines. Twooo. (Touch lines.) *3, 4, 5.*

i. Your turn to get it going and count the lines.
- (Point to **2.**) How many are in this group? (Signal.) *2.*
- Get it going. *Twooo.* (Touch lines.) *3, 4, 5.*
  (Repeat until firm.)
- How many are in both groups? (Signal.) *5.*
  Yes, 2 plus 3 lines equals 5.

---

from Lesson 26, Exercise 4

In steps A through D, children say problems and indicate how many each problem tells them to plus. 2 plus 7 tells children to plus 7.

The teacher then asks, "How do you plus 7?"

Children respond, "Make 7 lines."

In steps E through I, children read 2 plus 3, indicate how many the problem tells them to plus, and answer the question, "How do I plus 3?"

The teacher makes lines. Children count the lines and tell the teacher when to stop making them.

After making the lines, the teacher touches 2. The children get 2 going and count for each line as the teacher touches them. The teacher repeats this counting routine until children can count accurately and on time with what the teacher touches.

Teaching Note: Before this lesson, children have practiced all of the component skills solving these types of addition problems. Children have also answered the question "How do you plus ___?" (*Make ___ lines.*) This is the first time, however, that children have answered this question in the context of a problem-solving routine.

It's a good idea to repeat the routine for each problem at least once. The reason is that you want to make it clear to the children that they count the lines two different ways—first as a group of 3 lines, then as 3, 4, 5. This is not something lower performers learn without adequate practice. For lower performers, repeat the routine presented in the exercise.

If you are using the Board Displays, you will display in step E [26:4A]: 2 + 3. In step G, however, you will show 3 build-up displays, [26:4B1], [26:4B2], and [26:4B3].

On later lessons, the routine the teacher presents has fewer steps, and children are responsible for more details of the routine. On Lesson 34 the children work two problems in their Workbook.

Following is the part of the exercise that presents the first problem:

EXERCISE 10: ADDITION

a. Touch the snake on your worksheet. ✔
(Teacher reference:)

🐍 5 + 2 = | 6 + 3 =

The problems next to the snake are plus problems. For each problem, you'll work the whole thing. You'll figure out the number in both groups. Then you'll write the number on the other side of the equals.
• Touch the first problem. ✔
• Touch and read the problem. Get ready. (Tap 4.) *5 plus 2 equals.*
b. Touch the number you will make lines for. ✔
• What number are you touching? (Signal.) *2.*
c. I'll tap. You'll count to yourself and make lines under the 2. Tell me when to stop. Get ready. (Tap 2.) (Children make lines.)
(Teacher reference:)

🐍 5 + 2 = | 6 + 3 =
‖

d. Now you'll count both groups.
• Touch the number you'll get going. ✔
• Get it going. *Fiiive.* Touch and count. (Tap 2.) *6, 7.*
(Repeat until firm.)
• How many lines in both groups? (Signal.) *7.*
• So what do you write on the other side of the equals? (Signal.) *7.*
e. (Write on the board:)   [34:10A]

7

• (Point to 7.) Write 7 on the other side of the equals. Make your 7 look like this 7.
(Observe children and give feedback.)
(Write to show:)   [34:10B]

5 + 2 = 7
‖

f. Here's what you should have.
• Are the sides equal? (Signal.) *Yes.*
Yes, there are 7 on both sides.
• How many are on both sides? (Signal.) *7.*
g. Touch and read the problem and the answer. Get ready. (Tap 5.) *5 plus 2 equals 7.*
(Repeat step g until firm.)

from Lesson 34, Exercise 10

In step A, children read the problem. In step B, they identify the numbers they will make lines for. In step C, children make 2 lines under the 2. In steps D and E, children touch 5, get it going, touch and count on for the lines (6, 7), then write 7 on the other side of the equals. In step G, children touch the symbols as they read the problem and the answer.

Teaching Note: You should be able to present the routine quite fast because children should be well practiced in each component of the procedure.

Note: This is the same series of steps children will apply to problems that add two-digit numbers. The only difference is that for the two-digit problems children express the

number they plus as Ts and lines, not just lines. In other words, before the end of the program, the steps are systematically eliminated so children can work any addition problem with a two-digit sum without assistance. However, at each step of the process, children should perform accurately.

On Lesson 48, children review the steps for working each problem in a set. Then children work each problem in the set with almost all of the prompts eliminated.

Here's part of the exercise from Lesson 48:

• Your turn: Touch and read the first problem next to the car. Get ready. (Tap 4.) *17 plus 2 equals.*
(Repeat until firm.)
b. Touch and read the next problem. Get ready. (Tap 4.) *1 plus 4 equals.*
• Touch and read the last problem. Get ready. (Tap 4.) *13 plus 5 equals.*
• Touch and read the first problem again. Get ready. (Tap 4.) *17 plus 2 equals.*
(Repeat step b until firm.)
c. For each problem, you'll make lines under one of the numbers. Then you'll count for both groups and write the number.
• Touch the number you'll make lines for in the first problem. ✔
• How many lines will you make? (Signal.) *2.*
• Touch the number you'll get going. ✔
• What number will you get going? (Signal.) *17.*
d. Touch the number you'll make lines for in the next problem. ✔
• How many lines will you make? (Signal.) *4.*
• Touch the number you'll get going. ✔
• What number will you get going? (Signal.) *1.*
e. Touch the number you'll make lines for in the last problem. ✔
• How many lines will you make? (Signal.) *5.*
• Touch the number you'll get going. ✔
• What number will you get going? (Signal.) *13.*
(Repeat steps c through e until firm.)
f. Touch the number you'll make lines for in the **first** problem again. ✔
• Work the first problem. Make lines, count for both groups, and write the number to make the sides equal.
(Observe children and give feedback.)
(Teacher reference:)

🚗 17 + 2 = 19 | 1 + 4 =
‖
13 + 5 =

g. Check your work.
• Touch and read the whole thing. (Tap 5.) *17 plus 2 equals 19.*
• What does 17 plus 2 equal? (Signal.) *19.*

from Lesson 48, Exercise 7

Later in the program, children work plus problems entirely independently.

TURN-AROUNDS (COMMUTATIVE PROPERTY OF ADDITION)

CMC Level A introduces the commutative property of addition: 1 + 3 = 3 + 1. The program refers to these pairs as turn-arounds. "Say the turn-around for 1 + 3." (*3 + 1.*)

The Turn-Around track starts on Lesson 73. Work on turn-arounds continues to the end of the program.

1. First children say the turn-around for parts of addition equations: *7 + 1 = 1 + 7.*

2. Children say turn-arounds for addition equations: "Say the turn-around for 7 + 3 = 10." (*3 + 7 = 10.*)

3. Children work some unfamiliar problems by turning the problem around; for instance, 8 + 30. Children don't know the answer to this problem, but they do know the turn-around: 30 + 8. So they first say the equation for the turn-around, then they say the equation for the initial problem.
   (*30 + 8 = 38, so 8 + 30 = 38.*)

4. Children discriminate between equations that have legitimate turn-arounds and those that don't. "Can you say the turn-around for 50 + 3?" (*Yes.*) "Can you say the turn-around for 50 – 3?" (*No.*)

Note that most at-risk students require considerable practice before they are able to reliably say the phase-2 type turn-around for addition equations like 3 + 1 = 4. Saying turn-around equations is difficult because the plus, the equals, and the answer remain in the same place, while the first two numbers change positions. The program provides several days of practice on phase-1 type turn-around problems, which are less complicated than phase-2 type statements. The practice on saying the turn-around problems reduces confusion when children say the turn-around equations (1 + 3 = 4, 3 + 1 = 4). Understand that for at-risk children, you may have to repeat some turn-around examples several times before many children are able to reliably say them.

Once children have become facile performing tasks in phases 1 and 2, they should proceed quickly through phases 3 and 4.

Phase 1

Here's an early exercise from Lesson 73, in which children say the turn-around for parts of addition equations:

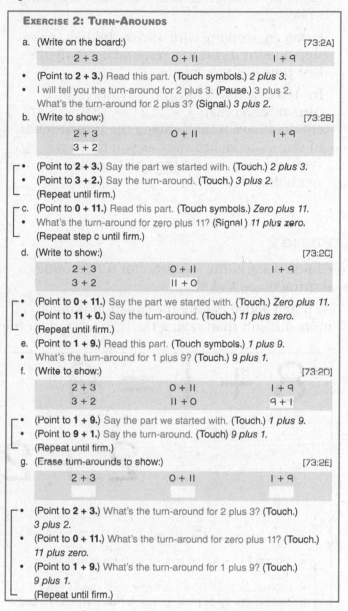

Lesson 73, Exercise 2

In step A, the teacher models the turn-around for 2 + 3.

In step B, the teacher writes the turn-around. Then children say the part they started with and the turn-around.

In steps C and D, children repeat what they did in steps A and B with a new problem, 0 + 11. However, this turn-around (0 + 11) is not modeled.

In steps E and F, children work with 1 + 9. In step G, the teacher erases the turn-arounds, and the children say the turn-around for each problem.

Teaching Note: In steps A and B, do not use a conversational tone. Say 2 + 3 and 3 + 2 in a very clear voice, at the rate children would be able to read them. This rate makes it much easier for children to hear what they are going to say.

If you are working with at-risk children, repeat step A, even if children respond correctly the first time it is presented.

In step G you will touch the symbols in turn-around order: 3, +, 2; 11, +, 0; 9, +, 1. If children have trouble saying the turn-arounds in step G, model the task as you touch the symbols. Then repeat the task touching the symbols.

Teacher: (Point to 0 + 11.) What's the turn-around for zero plus 11?
Some children: *Zero plus 11.*
Teacher: (Point to 0 + 11.) My turn to say the turn-around for (touch symbols) zero plus 11. (Touch symbols.) 11 plus zero. What's the turn-around for zero plus 11? (Touch symbols.)
Children: *11 plus zero.*

Phase 2

Children say entire equation for turn-arounds starting on Lesson 82.

As noted above, saying the equation is much more difficult than saying the turn-around for the part because children have to remember that the first part of the equation is turned around, but the last part remains the same.

Here's the introductory exercise:

$$8 + 1 = 9 \qquad 17 + 4 = 21$$
$$2 + 28 = 30$$

EXERCISE 3: TURN-AROUND EQUATIONS

a. (Point to **8 + 1 = 9.**) This equation is (Touch symbols.) 8 plus 1 equals 9.
• Say the turn-around for 8 plus 1. Get ready. (Tap 3.) *1 plus 8.*
b. (Point to **17 + 4 = 21.**) This equation is (Touch symbols.) 17 plus 4 equals 21.
• Say the turn-around for 17 plus 4. Get ready. (Tap 3.) *4 plus 17.*
c. (Point to **2 + 28 = 30.**) This equation is (Touch symbols.) 2 plus 28 equals 30.
• Say the turn-around for 2 plus 28. Get ready. (Tap 3.) *28 plus 2.*
(Repeat steps a through c until firm.)
d. (Point to **8 + 1 = 9.**) Let's figure out the turn-around equations.
• Say the turn-around for 8 plus 1. Get ready. (Tap 3.) *1 plus 8.*
• 8 plus 1 equals 9, so what does 1 plus 8 equal? (Signal.) *9.*
• Start with 1 and say the turn-around equation. Get ready. (Tap 5.) *1 plus 8 equals 9.*
(Repeat step d until firm.)

e. (Point to **17 + 4 = 21.**) Say the turn-around for 17 plus 4. Get ready. (Touch.) *4 plus 17.*
• 17 plus 4 equals 21, so what does 4 plus 17 equal? (Signal.) *21.*
• Start with 4 and say the turn-around equation. Get ready. (Tap 5.) *4 plus 17 equals 21.*
(Repeat step e until firm.)
f. (Point to **2 + 28 = 30.**) Say the turn-around for 2 plus 28. Get ready. (Touch.) *28 plus 2.*
• 2 plus 28 equals 30, so what does 28 plus 2 equal? (Signal.) *30.*
• Start with 28 and say the turn-around equation. Get ready. (Tap 5.) *28 plus 2 equals 30.*
(Repeat step f until firm.)

Lesson 82, Exercise 3

Connecting Math Concepts

In steps A, B, and C the teacher reads the equations and children say the turn-arounds for the first part of each equation: 8 + 1, 17 + 4, and 2 + 28. In step D, children say the turn-around for 8 + 1. The teacher tells them what 8 + 1 equals and asks what 1 + 8 equals. Then children say the turn-around equation: *1 + 8 = 9.*

Steps E and F repeat the tasks in step D for the other two equations.

> **Teaching Note:** If children have trouble saying the turn-around equation for 8 + 1 = 9, repeat the task, then repeat the step. When children are firm on all the tasks related to saying the turn-around for 8 + 1 = 9, they will have a model for what they will do with other equations. After children perform step D without errors, present step E immediately after. Reinforce children when they perform all of the tasks in the step without a mistake.

On the following lessons, the program systematically reduces the prompting so children become more independent saying turn-around equations. Later, addition equations are presented in the Workbook. Children write the turn-around equations. Here's the Workbook part of the exercise from Lesson 86:

---

c. Touch the equation 2 plus 4 equals 6 on worksheet 86. ✔

  2+4=6 | 5+24=29

• Say the turn-around equation for 2 plus 4. Get ready. (Tap 5.) *4 plus 2 equals 6.*

d. Touch the space where you'll write 4 plus 2 equals 6. (Observe children and give feedback.)

• What equation will you write in the space you're touching? (Tap 5.) *4 plus 2 equals 6.*

• Write it. (Observe children and give feedback.) (Teacher reference:)

  2+4=6   5+24=29
  4+2=6 _____

e. Touch the equation 5 plus 24 equals 29. ✔

• Say the turn-around equation for 5 plus 24. Get ready. (Tap 5.) *24 plus 5 equals 29.*

f. Touch the space where you'll write 24 plus 5 equals 29. (Observe children and give feedback.)

• What equation will you write in the space you're touching? (Tap 5.) *24 plus 5 equals 29.*

• Write it. (Observe children and give feedback.) (Answer key:)

  2+4=6 | 5+24=29
  4+2=6 | 24+5=29

from Lesson 86, Exercise 7

By the end of the program, children have learned to solve a variety of addition problems. Children write turn-around equations for all the types of familiar addition.

Phase 3

Phase 3 starts on Lesson 87. The examples children work are turn-arounds of familiar facts, such as 5 + 90. Children say the turn-around problem (*90 + 5*), which is a familiar problem. Then they are able to say the equation for the original problem.

Children will apply the procedure used to analyze turn-arounds to problems through the end of *CMC Level A*. They will learn to test unfamiliar problems to see if the turn-around is a familiar problem. If it is, they will be able to say the answer without calculating it.

Here's the first part of the exercise from Lesson 87:

$$5 + 90 =$$

$$1 + 34 =$$

$$0 + 50 =$$

$$8 + 70 =$$

EXERCISE 4: TURN-AROUND EQUATIONS

a. (Display page and point to problems.)   [87:4A]
   You don't know the answer to these problems. You do know the answers to the turn-arounds. You can figure out the answer to these problems by saying the turn-around equations.
• (Point to **5 + 90**.) Tell me the turn-around for (Touch.) 5 plus 90. Get ready. (Touch symbols.) *90 plus 5.*
• What does 90 plus 5 equal? (Signal.) *95.*
   So what does 5 plus 90 equal? (Signal.) *95.*
• Yes, 5 plus 90 equals 95. Say the equation for 5 plus 90. (Tap 5.) *5 plus 90 equals 95.*
b. (Point to **1 + 34**.) Tell me the turn-around for (Touch.) 1 plus 34. Get ready. (Touch symbols.) *34 plus 1.*
• What does 34 plus 1 equal? (Signal.) *35.*
   So what does 1 plus 34 equal? (Signal.) *35.*
• Say the equation for 1 plus 34. (Tap 5.) *1 plus 34 equals 35.*
   (Repeat step b until firm.)

c. (Point to **0 + 50**.) Tell me the turn-around for (Touch.) zero plus 50. Get ready. (Touch symbols.) *50 plus zero.*
• What does 50 plus zero equal? (Signal.) *50.*
• So what does zero plus 50 equal? (Signal.) *50.*
• Say the equation for zero plus 50. (Tap 5.) *Zero plus 50 equals 50.*
   (Repeat step c until firm.)
d. (Point to **8 + 70**.) Tell me the turn-around for (Touch.) 8 plus 70. Get ready. (Touch symbols.) *70 plus 8.*
• What does 70 plus 8 equal? (Signal.) *78.*
   So what does 8 plus 70 equal? (Signal.) *78.*
• Say the equation for 8 plus 70. (Tap 5.) *8 plus 70 equals 78.*
   (Repeat step d until firm.)

60   Lesson 87   Connecting Math Concepts

from Lesson 87, Exercise 4

Teaching Note: This exercise provides a strategy children may use to figure out the answers to turn-arounds of place-value addition problems and problems that add 1 or zero.

Phase 4

Phase 4 begins on Lesson 93 after children have worked on turn-arounds for 20 lessons. Phase 4 shows children that you can turn around addition problems, but subtraction problems cannot be turned around.

Here's the exercise from Lesson 93:

EXERCISE 5: TURN-AROUND DISCRIMINATION

a. You've learned that you can turn around plus problems. You cannot turn around take-away problems.
• Can you turn around take-away problems? (Signal.) *No.*
   (Repeat step a until firm.)
b. (Write on the board)   [93:5A]

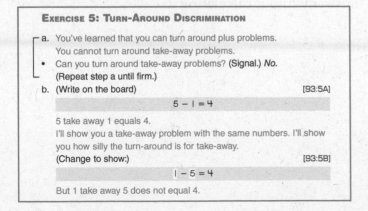

$$5 - 1 = 4$$

5 take away 1 equals 4.
I'll show you a take-away problem with the same numbers. I'll show you how silly the turn-around is for take-away.
(Change to show:)   [93:5B]

$$1 - 5 = 4$$

But 1 take away 5 does not equal 4.

c. We start with 1 line.
   (Write to show:)   [93:5C]

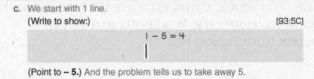

$$1 - 5 = 4$$

(Point to **– 5**.) And the problem tells us to take away 5. That's silly because we can't do that because we don't have 5 lines.
• How many lines do we have ? (Signal.) *1.*
   (Erase board.)
d. Here's the rule: You can't turn around problems that take away.
• Can you turn around problems that take away? (Signal.) *No.*
• Can you turn around problems that plus? (Signal.) *Yes.*
   (Repeat step d until firm.)
e. I'll say problems. Tell me if you can turn them around.
• Listen: 5 plus 2. Say the problem. Get ready. (Tap 3.) *5 plus 2.*
• Can you turn around 5 **plus** 2? (Signal.) *Yes.*
• Say the turn-around for 5 plus 2. Get ready. (Tap 3.) *2 plus 5.*
   (Repeat step e until firm.)
f. Listen: 5 **take away** 4. Say the problem. Get ready. (Tap 3.) *5 take away 4.*
• Can you turn around 5 take away 4? (Signal.) *No.*
   (Repeat step f until firm.)
g. Listen: 10 **take away** 1. Say the problem. Get ready. (Tap 3.) *10 take away 1.*
• Can you turn around 10 take away 1? (Signal.) *No.*
   (Repeat step g until firm.)
h. Listen: 4 **plus** 3. Say the problem. Get ready. (Tap 3.) *4 plus 3.*
• Can you turn around 4 **plus** 3? (Signal.) *Yes.*
• Say the turn-around for 4 plus 3. Get ready. (Tap 3.) *3 plus 4.*
   (Repeat step h until firm.)
i. Listen: 4 **take away** 1. Say the problem. Get ready. (Tap 3.) *4 take away 1.*
• Can you turn around 4 take away 1? (Signal.) *No.*
   (Repeat step i until firm.)

Lesson 93, Exercise 5

Step A models and tests the rule that you can't turn around take-away problems. In steps B and C, the teacher shows how silly the turn-around equation is for problems that take away. In steps E through I, the teacher says a problem that pluses or a problem that takes away. Children repeat the problem, then they apply the addition rule to each problem to determine if it can be turned around. If the problem can be turned around, children say the turn-around.

Teaching Note: In step D, model the idea that turn-arounds for take-away problems are perfectly ridiculous because the turn-around takes away more than you have.

Apply the rule if children do have trouble:

Example:

Teacher: Can you turn around 5 take away 4?

Some children: *Yes.*

Use the following pair of questions to firm the discrimination.

Teacher: Does 5 take away 4 plus?

Children: *No.*

Teacher: So can you turn around 5 take away 4?

Children: *No.*

This pair of parallel questions, "Does ___ plus? So can you turn around ___?" works for discriminating all plus and take-away examples.

On Lesson 95, children combine phases 2 and 4. The teacher presents addition and take-away equations on the board. Children determine whether each equation can be turned around. For each equation that can be turned around, children say the turn-around equation and the teacher writes it.

Here's part of the exercise:

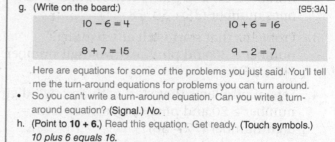

g. (Write on the board:) [95:3A]

10 − 6 = 4	10 + 6 = 16
8 + 7 = 15	9 − 2 = 7

Here are equations for some of the problems you just said. You'll tell me the turn-around equations for problems you can turn around.
• So you can't write a turn-around equation. Can you write a turn-around equation? (Signal.) *No.*

h. (Point to **10 + 6**.) Read this equation. Get ready. (Touch symbols.) *10 plus 6 equals 16.*
• Can you turn around 10 plus 6? (Signal.) *Yes.*
• Raise your hand when you can tell me the turn-around equation for 10 plus 6 equals 16. ✔
• Everybody, say the turn-around equation. Get ready. (Tap 5.) *6 plus 10 equals 16.*
(Write to show:) [95:3B]

10 − 6 = 4	10 + 6 = 16
	6 + 10 = 16
8 + 7 = 15	9 − 2 = 7

i. (Point to **8 + 7**.) Read this equation. Get ready. (Touch symbols.) *8 plus 7 equals 15.*
• Can you turn around 8 plus 7? (Signal.) *Yes.*
• Raise your hand when you can tell me the turn-around equation for 8 plus 7 equals 15. ✔
• Everybody, say the turn-around equation. Get ready. (Tap 5.) *7 plus 8 equals 15.*
(Write to show:) [95:3C]

10 − 6 = 4	10 + 6 = 16
	6 + 10 = 16
8 + 7 = 15	9 − 2 = 7
7 + 8 = 15	

from Lesson 95, Exercise 3

Later children apply the turn-around strategy to two-digit problems such as 54 + 34, 54 − 34, and 54 + □ = 60. Children complete the equations. Then, for problems that plus, children write the turn-around equation below.

TWO-DIGIT ADDITION PROBLEMS

The first problems children work that add two-part numbers start on Lesson 75. Work on adding two-digit values continues to the end of the program. All these problems have two addends and sums less than 100; however, the skills children learn in Level A enable them to solve problems with sums over 100.

Two-digit addition in *CMC Level A* focuses on five types of problem sets:

1. Problems that start with any two-digit number ≥ 20 and plus a tens number (35 + 50)

2. Sets of problems that start with any two-digit number ≥ 20 and plus a tens number or a ones number (35 + 50, 78 + 1)

3. Problems that start with any two-digit number ≥ 20 and plus any two-digit number (47 + 25)

4. Sets of problems that start with any two-digit number ≥ 20 and plus or take away a two-digit number (77 − 56)

5. Mixed sets of addition and subtraction problems containing at least one two-digit number

Problem Set 1

Starting on Lesson 75, children apply this count-on skill to plusing tens numbers (decade values).

Here's part of the exercise from Lesson 81:

EXERCISE 7: ADDING TENS

a. Start with 20 and count by tens to 90. Get 20 going. *Twentyyy.*
Count. (Tap 7.) *30, 40, 50, 60, 70, 80, 90.*
(Repeat step a until firm.)

b. Now you'll start with 25 and plus tens to 95. What number will you start with? (Signal.) *25.*

• Get 25 going. *Twenty-fiiive.* Plus tens. (Tap 7.) *35, 45, 55, 65, 75, 85, 95.*
(Repeat until firm.)

c. Now you'll start with 23 and plus tens to 93. What number will you start with? (Signal.) *23.*

• Get 23 going. *Twenty-threee.* Plus tens. (Tap 7.) *33, 43, 53, 63, 73, 83, 93.*
(Repeat until firm.)

d. (Write on the board:)   [81:7A]

$$35 + \boxed{}$$
$$\text{T T T T}$$

You're going to work this problem.

• (Point to **35**.) What number? (Touch.) *35.*
• (Point to **T**.) What number do you plus for each T? (Touch.) *10.*

e. (Point to **35**.) Get it going. (Touch **35**.) *Thirty-fiiive.* (Touch **Ts**.) *45, 55, 65, 75.*
(Repeat until firm.)

• How many in both groups? (Signal.) *75.*

f. (Write to show:)   [81:7B]

$$35 + \boxed{} = 75$$
$$\text{T T T T}$$

• (Point to **TTTT**.) Count for the Ts to figure out the number that goes in the box. Get ready. (Touch.) *10, 20, 30, 40.*
• What number do I write in the box? (Signal.) *40.*
(Write to show:)   [81:7C]

$$35 + \boxed{40} = 75$$
$$\text{T T T T}$$

• (Point to **35**.) Read the equation. Get ready. (Touch symbols.) *35 plus 40 equals 75.*
(Erase board.)

---

from Lesson 81, Exercise 7

In step A, children start with 20 and count by tens to 90. In step B, they start with 25 and plus tens to 95. In step C, children start with 23 and plus tens to 93.

The problem presented in steps D, E, and F starts with the number 35 and pluses four Ts. Children get 35 going and plus ten for each T. The teacher writes = **75**. Then children count for the Ts to figure out the addend number that goes in the box. The teacher completes the equation, 35 + **40** = 75, and children read it.

In the remaining steps of the exercise (not shown here) children work the same problem in their Workbook.

Teaching Note: Repeat the counting for steps A, B, and C until children are completely firm, especially step B. If children are not firm on this counting, they will have trouble working the problem.

In step E, if children keep counting after 75, say stop. Repeat the counting at a slower pace. Make touching each T more noticeable. Praise them for stopping after the last T.

Later, children work problems that show two-digit numbers plus a decade number. Children follow the strategy for plusing that they've learned and make the Ts for the number that's plused. Then they get the number the problem starts with going, touch and count for the Ts, and complete the equation.

Here's the set of Workbook problems from
Lesson 84:

$$56+40=\boxed{}\quad\Big|\quad 34+30=\boxed{}$$

Lesson 84, Workbook

Problem Set 2

On Lesson 86 (Exercise 4) children complete
Workbook problems that plus either single-digit
or two-digit numbers: 67 + 20, 67 + 4, and 24 +
40. In the first part of the exercise, children read
each problem and make either Ts or lines for the
number that is plused.

Here's the next part of the exercise:

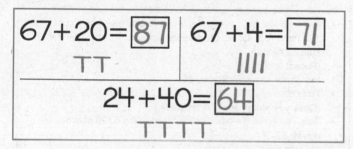

Lesson 86, Answer Key

---

j. Touch the problem 67 plus 20 again. ✔
    You'll get a number going and touch and count for the Ts to figure
    out the answer. How many do you plus for each T? (Signal.) *10.*
•   Touch the number you'll get going. ✔
•   Get it going. *Sixty-sevennn.* Count for the Ts. (Tap 2.) *77, 87.*
    (Repeat until firm.)
•   What's the answer? (Signal.) *87.*

k. Touch the problem 67 plus 4. ✔
    You'll get a number going and touch and count for the lines to figure
    out the answer. How many do you plus for each line? (Signal.) *1.*
•   Touch the number you'll get going. ✔
•   Get it going. *Sixty-sevennn.* Count for the lines. (Tap 4.)
    68, 69, 70, 71.
    (Repeat until firm.)
•   What's the answer? (Signal.) *71.*

l. Touch the problem 24 plus 40. ✔
    You'll get a number going and touch and count for the Ts to figure
    out the answer. How many do you plus for each T? (Signal.) *10.*
•   Touch the number you'll get going. ✔
•   Get it going. *Twenty-fouuur.* Count for the Ts. (Tap 4.) *34, 44, 54, 64.*
    (Repeat until firm.)
•   What's the answer? (Signal.) *64.*

from Lesson 86, Exercise 4

For each problem, children first answer "How
many do you plus for each ___?" (Top line.) Next,
children touch the number they'll get going, get it
going, and touch and count for each T or line.

> **Teaching Note:** The most likely mistake
> children will make is saying the wrong number
> when they touch the first counter below the
> number that's plused. They may have trouble
> with 67 + 20 and say 68 when they touch the
> first T.
>
> Be sensitive to children's response to the first T
> or line. If they hesitate or make a mistake, stop
> them immediately and say the correct number.
> Then direct them to do the counting again.
>
> If children make mistakes on both 67 + 20
> **and** 67 + 4, first correct each problem, then
> repeat the pair in sequence so children see the
> difference.

Problem Set 3

After children practice problem sets of phase 2 for
a few days, children work problems that plus two-
digit numbers that have both Ts and lines.

These problem sets are first presented on Lesson
90. In the first part of the exercise, children
practice assorted counting tasks that plus ones or
plus tens. Included in these counting tasks are the
counting tasks children use to solve the problems
presented later in the exercise.

Here's part of the introductory exercise that
directs the work for the first problem,
41 + 35:

c. (Write on the board:)   [90:5A]

$$41 + 35 =$$
TTTIIIII

- (Point to **41**.) Read this problem. Get ready. (Touch symbols.)
  41 plus 35 equals.
  I've already made the Ts and lines for 35. I'm going to count for both groups. I'll get 41 going and touch and count for the Ts and lines.
  (Touch **41**.) Forty-wuuun. (Touch Ts.) 51, 61, seventy-wuuun. (Touch lines.) 72, 73, 74, 75, 76.
- (Point to **41**.) Your turn to count for both groups. Get it going.
  (Touch **41**.) *Forty-wuuun.* Count for the Ts. (Touch Ts.) *51, 61, seventy-wuuun.* Count for the lines. (Touch lines.) *72, 73, 74, 75, 76.*
  (Repeat until firm.)
- What's the answer? (Signal.) *76.*
  (Write to show:)   [90:5B]

$$41 + 35 = 76$$
TTTIIIII

- (Point to **41**.) Read the equation. Get ready. (Touch symbols.)
  41 plus 35 equals 76.

from Lesson 90, Exercise 5

Teaching Note: Before presenting this exercise, practice the touching and counting at the rate you expect children to do the counting. In the first part of the step, the teacher models touching the counters and counting. Remember to hold the number you get going, 41, and the number for the final T, 71. Hold both numbers for 2 seconds. Touch 41 as long as you say "forty-wuuun" and "seventy-wuuun." Holding 71 is important because it is an alert that the next number does not plus ten, but rather pluses one.

In later lessons, you provide fewer models and less prompting as children's responsibility increases for making the Ts and lines and doing the touching and counting.

Here's part of the exercise from Lesson 95:

c. Touch the problem 31 plus 22 ☞
   equals. ✔
- You're going to make Ts and lines for one of the numbers. What number? (Signal.) 22.
- Touch the problem 28 plus 52 equals. ✔
- You're going to make Ts and lines for one of the numbers. What number? (Signal.) 52.
- Make the Ts and lines for the numbers in both problems.
  (Observe children and give feedback.)
  (Teacher reference:)

  31+22=   28+52=
  TTII   TTTTTII

d. Now you're going to figure out the answer for the first problem.
- Touch 31 and get it going. Get ready. *Thirty-wuuun.* Count for the Ts. (Tap 2.) *41, fifty-wuuun.* Count for the lines. (Tap 2.) *52, 53.*
  (Repeat until firm.)
- What's the answer? (Signal.) *53.*
- Write 53.
  (Observe children and give feedback.)
- Touch and read the equation. Get ready. (Tap 5.) *31 plus 22 equals 53.*
e. Now you're going to figure out the answer for the next problem.
- Touch 28 and get it going. Get ready. *Twenty-eieieight.* Count for the Ts. (Tap 5.) *38, 48, 58, 68, seventy-eieieight.* Count for the lines. (Tap 2.) *79, 80.*
  (Repeat until firm.)
- What's the answer? (Signal.) *80.*
- Write 80.
  (Observe children and give feedback.)
- Touch and read the equation. Get ready. (Tap 5.) *28 plus 52 equals 80.*
  (Answer key:)

  31+22= 53   28+52=80
  TTII   TTTTTII

from Lesson 95, Exercise 6

In step C, children make the Ts and lines for the number that's plused in both problems. In step D, children work the first problem. In step E, children work the other problem.

Teaching Note: In step E, children work a problem (28 + 52) that would traditionally require carrying. The strategy children apply is the same one they use for problems that don't require carrying. Children make Ts and lines for 52, then get 28 going and count on for the Ts and the lines.

In step D, direct children to do the touching and counting two or three times. Attend to whether individual children are touching the appropriate counters and are saying the numbers. If you have doubts, call on several children to do the counting individually.

Problem Sets 4 and 5

These problem sets are not introduced until children have mastered subtracting two-digit numbers (see the Take Away track, which follows this Plus track) and after children are well practiced in adding tens to one-digit and two-digit numbers.

Here's a problem set of type 4 from Lesson 114 and a problem set of type 5 from Lesson 115:

$$54-23=$$

$$57+23=$$

$$\begin{array}{r} 32 \\ -\ 12 \\ \hline \end{array}$$

Lesson 114, Workbook

$$\begin{array}{r} 44 \\ -\ 12 \\ \hline \end{array}$$

$$\begin{array}{r} 5 \\ +\ 26 \\ \hline \end{array}$$

$$36-30=$$

Lesson 115, Workbook

COLUMN PROBLEMS: ADDITION

Note that some of the problems shown are written in columns. In Lesson 85, children are introduced to column problems. A demonstration shows that column problems have symbols that correspond to row problems. All the symbols are the same except for =, which becomes a single line, a bar, in column problems.

Here's part of the exercise that introduces column problems:

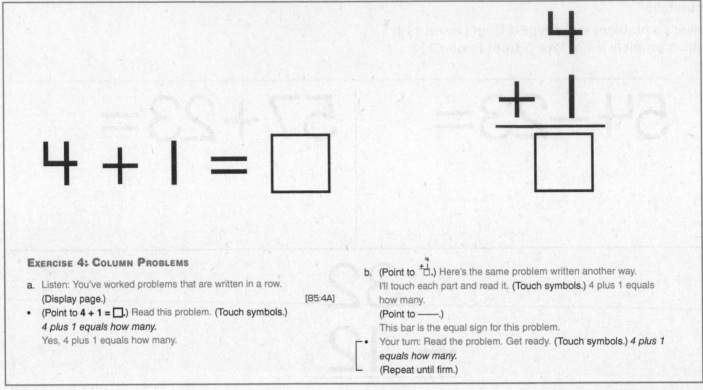

EXERCISE 4: COLUMN PROBLEMS

a. Listen: You've worked problems that are written in a row. (Display page.)   [85:4A]
• (Point to **4 + 1 = ☐**.) Read this problem. (Touch symbols.) *4 plus 1 equals how many.*
Yes, 4 plus 1 equals how many.

b. (Point to ⁴₊₁☐.) Here's the same problem written another way. I'll touch each part and read it. (Touch symbols.) 4 plus 1 equals how many.
(Point to ——.)
This bar is the equal sign for this problem.
• Your turn: Read the problem. Get ready. (Touch symbols.) *4 plus 1 equals how many.*
(Repeat until firm.)

from Lesson 85, Exercise 4

In step A, children read a familiar row problem. In step B, the teacher touches each symbol of the column problem as children read them.

Teaching Note: It's important to touch each symbol as you or the children read. Make sure children touch the bar in the column problem and identify it as equals. Adjust the rate at which you read the problem so you can coordinate touching the symbols with what you say.

Children work the first column problems that add pairs of two-digit numbers on Lesson 96. The strategy for solving the problems that the teacher presents on the board is identical to the strategy children have used to solve problems that add two-digit numbers written in rows. The difference is that the counters are made to the right of the number that's plused.

On the following lessons, children work column problems, such as the following: ⁵₊₀ ⁹₊₁ ⁶⁰₊₄ ⁶⁰₊₈.

These exercises ensure that children write the digits of the answers in the correct columns.

Here's part of the exercise from Lesson 100:

f. Touch the column problem 43 plus 24 equals how many on worksheet 100. ✔
(Teacher reference:)

$$\begin{array}{r} 43 \\ + 24 \\ \hline \square \end{array}$$

• You'll make Ts and lines for one of the numbers. Which number? (Signal.) *24.*
• How many Ts do you make for 24? (Signal.) *2.*
• How many lines do you make for 24? (Signal.) *4.*
• There is no line to show you where to make Ts and lines. Touch where you'll make them.
(Observe children and give feedback.)
• Make 2 Ts and 4 lines next to 24. Then count for both groups and write the answer. Put your pencil down when you've completed the equation.
(Observe children and give feedback.)
(Answer key:)

$$\begin{array}{r} 43 \\ + 24 \quad \text{TTIIII} \\ \hline \boxed{67} \end{array}$$

• Touch and read the equation. Get ready. (Tap 5.) *43 plus 24 equals 67.*

from Lesson 100, Exercise 7

Children indicate that they will make Ts and lines for 24. They say how many Ts and how many lines they'll make. Then children make the counters, solve the problem, and write the answer.

Teaching Note: In *CMC Level A*, the only difference between solving column problems and solving problems written in rows is where children make counters. If children are firm working two-digit addition problems written in rows, children should be able to work the two-digit column problems with little trouble.

Expect some children to have trouble unless they verbalize the counting. Show them how to touch and count "quietly" to themselves. Don't require them to whisper, just count quietly. Praise children for not counting in a big voice.

If quite a few children have trouble working the problem, first check to make sure that each child has made the correct number of Ts and lines. Then direct the group to touch and count out loud (*Forty-threee, 53, sixty-threee, 64, 65, 66, 67*). After the group is firm counting out loud, repeat the step as it is specified in the script.

ADDITION FACTS

CMC Level A teaches the most basic facts. Facts are problems that children are expected to solve without performing counting or calculations. Children have mastered facts when they solve them without pausing to figure out the answer.

The categories of addition facts taught in the program are:

• Plus-1 Facts
• Plus-0 Facts (Additive Identity)
• Plus-2 Facts
• Plus-10 Facts
• Teen-Number Facts (Place Value—Turn-Arounds—Plus 10)

All plus-0, plus-1, place-value and plus-10 facts, and turn-around facts apply to numbers 1–99. The plus-2 facts have a more limited range.

Plus-1 Facts

Plus-1 facts are introduced first because they are so closely related to counting and saying the next number. On Lesson 42, plus-1 facts are first introduced on a number line. Starting on Lesson 44, the teacher says, "Listen: 5. What number? What's the next number?" After a correct response, the teacher asks, "So what does 5 + 1 equal?"

Later, when children are answering plus-1 problems without prompting, use the next-number correction if they make a mistake. If children read the problem 39 + 1 and say that the answer is 30, correct by saying, "Listen: 37, 38, thirty-niiine. What's the next number?" "So what's 39 + 1?" Then repeat the problem after children finish the set of problems.

On Lesson 42, the teacher presents plus-1 problems without connecting them to a next-number task.

Here's the exercise:

EXERCISE 6: PLUS 1

Note: (Do not display page until step b.)

a. Listen: Here's a rule about plusing 1. When you **plus 1,** you say the next number.
Listen again: When you **plus 1,** you say the next number.
• What do you do when you **plus 1?** (Signal.) *Say the next number.*
(Repeat until firm.)

b. (Display page and point to number line.) [42:6A]
For each task, I'll say **plus 1 equals.**
When I say **plus 1 equals,** you'll say the next number.
Do it with me. (Point to **1.**) (Respond with students.)
• What number? (Touch.) *1.*
• Plus 1 equals? (Touch **2.**) *2.*
• Plus 1 equals? (Touch **3.**) *3.*
• Plus 1 equals? (Touch **4.**) *4.*
• Plus 1 equals? (Touch **5.**) *5.*
(Repeat step b until firm.)

c. All by yourself: (Do not respond with children.)
• (Point to **1.**) What number? (Touch.) *1.*
• Plus 1 equals? (Touch **2.**) *2.*
Yes, 2.
• Plus 1 equals? (Touch **3.**) *3.*
Yes, 3.
• Plus 1 equals? (Touch **4.**) *4.*
Yes, 4.
• Plus 1 equals? (Touch **5.**) *5.*
Yes, 5.
(Repeat step c until firm.)

d. This time I'll mix it up.
• (Point to **4.**) What number? (Touch.) *4.*
• Plus 1 equals? (Touch **5.**) *5.*
Yes, 4 plus 1 equals 5.
• Everybody, say that. Get ready. (Signal.) *4 plus 1 equals 5.*
(Repeat step d until firm.)

e. (Point to **2.**) What number? (Touch.) *2.*
• Plus 1 equals? (Touch **3.**) *3.*
Yes, 2 plus 1 equals 3.
• Everybody, say that. Get ready. (Signal.) *2 plus 1 equals 3.*

Lesson 42, Exercise 6

Teaching Note: Remember to touch each next number as you or the children say it. You should present the tasks in step B at a reasonably brisk pace. Step C should be presented at the same pace. Do not talk much more slowly than you normally talk. You can say the words "plus 1 equals" very quickly.

If children make a mistake in step C, point to the answer and say the answer. Then tell them, "Remember, the answer is the next number."

Repeat the task. Your goal should be that children respond correctly to all the tasks in step C before you proceed to step D.

In step D, point to 4 for a second before you ask "What number?"

After children have practiced saying plus-1 facts that are prompted by the number line, children say facts without the number-line prompt.

On Lesson 56, children review some of the teen plus-1 facts they are learning and then write answers to some problems in their Workbook.

Here's the exercise:

EXERCISE 7: PLUS 1

a. Here's the plus-1 equation that starts with 12.
12 plus 1 equals 13.
- Everybody, say the plus-1 equation that starts with 12. Get ready. (Signal.) *12 plus 1 equals 13.*
- Say the plus-1 equation that starts with 10. Get ready. (Signal.) *10 plus 1 equals 11.*
- Say the plus-1 equation that starts with 9. Get ready. (Signal.) *9 plus 1 equals 10.*
- Say the plus-1 equation that starts with 6. Get ready. (Signal.) *6 plus 1 equals 7.*
- Say the plus-1 equation that starts with 17. Get ready. (Signal.) *17 plus 1 equals 18.*
- Say the plus-1 equation that starts with 13. Get ready. (Signal.) *13 plus 1 equals 14.*
- Say the plus-1 equation that starts with 8. Get ready. (Signal.) *8 plus 1 equals 9.*
(Repeat tasks that were not firm.)
b. (Open workbooks to Lesson 56 and distribute to children.)
- (Hold up worksheet side 1 and point to **8 + 1 =**.) ☞
- Touch the plus-1 problems on your worksheet. ✔
- Touch and read the first problem. Get ready. (Tap 4.) *8 plus 1 equals.*
What's the answer? (Signal.) *9.*
- Touch and read the next problem. Get ready. (Signal.) *10 plus 1 equals.*
What's the answer? (Signal.) *11.*
c. Write answers to all the problems. Put your pencil down when you're finished.
(Observe children and give feedback.)
d. Check your work.
- Touch and read the first equation. Get ready. (Signal.) *8 plus 1 equals 9.*
- Touch and read the next equation. Get ready. (Signal.) *10 plus 1 equals 11.*
- (Repeat for remaining equations.) *17 + 1 = 18; 6 + 1 = 7.*
(Answer key:)

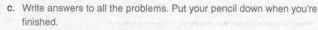

Lesson 56, Exercise 7

In step A, the teacher directs children to say plus-1 equations that start with different numbers. In step B, children are directed to their Workbook. Some of the problems are ones that they responded to in step A. Children touch and read each problem and say the answer. In step C, children write answers to all the problems.

Teaching Note: Make sure that children respond correctly to the verbal items in step A before presenting the Workbook practice with the same problems.

Children should not have difficulty with this kind of exercise. However, in step C, give feedback to children who are doing well and tell the class. "Nice writing, Nickie….I don't see children making any mistakes at all…!"

Plus-0 Facts (Additive Identity)

Plus-zero facts are introduced on Lesson 70 after children have learned many plus-1 facts. Plus-0 facts are very simple—the answer is the number in the problem other than zero—so introducing plus-zero facts too early may prompt poor strategies about how to learn facts.

Here's the exercise from Lesson 70:

EXERCISE 5: ZERO

a. You've worked problems that plus. You make 4 lines if you plus 4.
- How many lines do you make if you plus 4? (Signal.) *4.*
- How many lines do you make if you plus 3? (Signal.) *3.*
- How many lines do you make if you plus zero? (Signal.) *Zero.*
(Repeat until firm.)
b. (Write on the board:) [70:5A]

$$4 + 0 = \square$$
$$19 + 0 = \square$$

- (Point to **4 + 0 = □**.) Everybody, read this problem. Get ready. (Touch symbols.) *4 plus zero equals how many.*
(Repeat until firm.)
- How many are we starting with? (Signal.) *4.*
c. Now I'll show you how to make the lines when you plus zero. (Pause.) There, I plused zero lines. I didn't make any lines at all. So we end up with the same number we started with, 4.
(Write to show:) [70:5B]

$$4 + 0 = \boxed{4}$$
$$19 + 0 = \square$$

d. (Point to **4 + 0 = 4**.) I'll say the fact. (Touch symbols.) *4 plus zero equals 4.*
- Say the fact. Get ready. (Tap 5.) *4 plus zero equals 4.*
(Repeat step d until firm.)
e. (Point to **19 + 0 = □**.) Everybody, read the problem. Get ready. (Touch symbols.) *19 plus zero equals how many.*
- How many are we starting with? (Touch.) *19.*
- If we plus zero, how many do we end up with? (Signal.) *19.*
(Write to show:) [70:5C]

$$4 + 0 = \boxed{4}$$
$$19 + 0 = \boxed{19}$$

f. 19 plus zero equals 19. Say the fact. (Tap 5.) *19 plus zero equals 19.*
(Repeat step f until firm.)
g. You'll say more plus-zero facts.
- Listen: Say the fact for 11 plus zero. Get ready. (Signal.) *11 plus zero equals 11.*
Yes, 11 plus zero equals 11.
- Listen: Say the fact for 1 plus zero. Get ready. (Signal.) *1 plus zero equals 1.*
- Listen: Say the fact for 34 plus zero. Get ready. (Signal.) *34 plus zero equals 34.*
(Repeat step g until firm.)

Lesson 70, Exercise 5

In step A, children apply the rule they've learned about making lines and plusing to the number zero. Children know that zero refers to none or nothing. When they apply the rule for making lines for zero, they conclude that they don't make any lines. So they end up with the number they start with. In steps B through F, children work the problems 4 + 0 and 19 + 0.

In step G, children say facts for plus-zero problems that the teacher presents orally.

> **Teaching Note:** Treat adding zero as a simple trick. You don't plus anything, so you end up with the number you started with. Children should have no trouble with this exercise or later ones that plus zero.

On Lesson 73, children work mixed sets of problems that plus 1 and plus zero in their workbook.

Plus-2 Facts

Plus-2 facts are first introduced on Lesson 92, after children have worked extensively with plus-1 facts. The basic strategy for plus 2 is that it is the counting number **after** the number for plus 1. If 7 plus 1 equals 8, 7 plus 2 equals 9. Children must be very solid on their plus-1 facts before they begin instruction on plus-2 facts.

Here's the exercise from Lesson 92:

EXERCISE 5: PLUS 2

a. You're going to say answers to problems that plus 1. Then you'll say answers to problems that plus 2.
- Listen: 5 plus 1. What's 5 plus 1? (Signal.) *6.*
- So 5 plus 2 equals 7. What does 5 plus 2 equal? (Signal.) *7.*

b. Let's do those again. What does 5 plus 1 equal? (Signal.) *6.*
- What does 5 plus 2 equal? (Signal.) *7.*
- (Repeat step b until firm.)

c. New problem: 3 plus 1. What's 3 plus 1? (Signal.) *4.*
- So 3 plus 2 equals 5. What does 3 plus 2 equal? (Signal.) *5.*

d. Let's do those again. What does 3 plus 1 equal? (Signal.) *4.*
- What does 3 plus 2 equal? (Signal.) *5.*
- (Repeat step d until firm.)

e. New problem: 8 plus 1. What's 8 plus 1? (Signal.) *9.*
- So 8 plus 2 equals 10. What does 8 plus 2 equal? (Signal.) *10.*

f. Let's do those again. What does 8 plus 1 equal? (Signal.) *9.*
- What does 8 plus 2 equal? (Signal.) *10.*
- (Repeat step f until firm.)

g. I'm going to say each of the problems again. You'll tell me the answers.
- 5 plus 1. What does 5 plus 1 equal? (Signal.) *6.*
  So what does 5 plus 2 equal? (Signal.) *7.*
- 3 plus 1. What does 3 plus 1 equal? (Signal.) *4.*
  So what does 3 plus 2 equal? (Signal.) *5.*
- 8 plus 1. What does 8 plus 1 equal? (Signal.) *9.*
  So what does 8 plus 2 equal? (Signal.) *10.*
- (Repeat step g until firm.)

—————— **INDIVIDUAL TURNS** ——————
(Call on individual children to answer the following questions.)

- What does 5 plus 1 equal? (Call on a child.) *6.*
  So what does 5 plus 2 equal? *7.*
- What does 8 plus 1 equal? (Call on a child.) *9.*
  So what does 8 plus 2 equal? *10.*

Lesson 92, Exercise 5

The teacher presents different pairs of facts that show the relationship between plus-1 and plus-2 facts. In steps A, C, and E, children answer the plus-1 question, and the teacher models the answer to the plus-2 question. In the following step, each pair is repeated, without the teacher model (steps B, D, and F). In step G, all three pairs from steps B, D, and F are repeated.

Teaching Note: The exercise repeats each pair several times because children need a fair amount of practice to learn the relationship and to be able to generalize the pattern to examples that have not been taught.

When presenting the pair, stress the number that is plused:

New problem: 7 + 1. What's 7 + **1?**

So what's 7 + **2?**

Repeat steps B, D, and F until children are firm. As with other patterns, keep the examples rhythmical, and don't hesitate to repeat examples. The more children hear the pair of instances, the faster they'll learn the pattern. Children should be able to respond to them quickly and correctly.

Present the tasks in step G with the same pacing and rhythm.

If children have trouble with a plus-2 fact in later lessons, use the strategy of first presenting the corresponding plus-1 problem to prompt the plus-2 fact. For instance, a child misses the item: 27 + 2. Correct by asking "What's 27 plus 1? So what's 27 plus 2?"

Plus-10 Facts

In Lesson 103, children do pairs of problems that plus 10. By this lesson children have worked extensively on plusing tens in the context of counting and of two-digit addition. These plus-10 skills are systematically extended to facts.

The first problem has a decade number plus 10. The other problem adds 10 to another number from the same decade.

Here's the exercise:

$$40 + 10 =$$
$$46 + 10 =$$
$$70 + 10 =$$
$$74 + 10 =$$

EXERCISE 9: FACTS—Plus 10

a. (Display page and point to problems.)   [103:9A]
  You're going to plus 10 to two-part numbers.
• (Point to **40.**) Read this problem. Get ready. (Touch.) *40 plus 10 equals.*
• What's 40 plus 10? (Signal.) *50.*
• (Point to **46.**) So what's (Touch.) 46 plus 10? (Signal.) *56.*
b. (Point to **70.**) Read this problem. Get ready. (Touch.) *70 plus 10 equals.*
• What's 70 plus 10? (Signal.) *80.*
• (Point to **74.**) So what's (Touch.) 74 plus 10? (Signal.) *84.*
  (Repeat steps a and b until firm.)
c. Now you'll say answers to some hard plus-10 problems again.
• (Point to **46.**) What's (Touch.) 46 plus 10? (Signal.) *56.*
• (Point to **74.**) What's (Touch.) 74 plus 10? (Signal.) *84.*
  (Repeat step c until firm.)

d. Touch the problem 74 plus 10 equals how many on worksheet 103. ✔
  (Teacher reference:)

  74+10=☐   46+10=☐

  These are the plus-10 problems we just worked. Write the answers to all the problems. Put your pencil down when you've completed both the equations.
  (Observe children and give feedback.)
  (Answer key:)

  74+10=84   46+10=56

e. Check your work.
• Touch and read the first equation. Get ready. (Tap 5.) *74 plus 10 equals 84.*
• Touch and read the second equation. Get ready. (Tap 5.) *46 plus 10 equals 56.*

Lesson 103, Exercise 9

In step A, children read the first problem in the pair and say the answer ("What's 40 plus 10?"). Then they say the answer to the paired problem ("What's 46 plus 10?"). In step B, children repeat the tasks in step A with another pair of problems.

In step C, children answer the questions about the second problem in each pair. In step D children write answers to these problems in their Workbooks.

Teaching Note: After children answer the question for the decade number correctly in steps A and B, quickly present the next task. Emphasize the ones digit of the number from the same decade.

What's 40 + 10? *50.*

So what's 4**6** + 10? *5**6**.*

Children should not have trouble with these problems because they have plused 10 to a variety of two-digit numbers. This exercise

suggests a strategy for learning plus-10 facts. (For the problem: 57 + 10, children say the fact for 50 + 10 to themselves, then the fact for 57 + 10.)

After children have practiced working with pairs of plus-10 facts for a few lessons, they work sets of problems that include plus-10 and other familiar facts (+ 1, + 2, + 0, place value).

Here's the set of problems children work on Lesson 105:

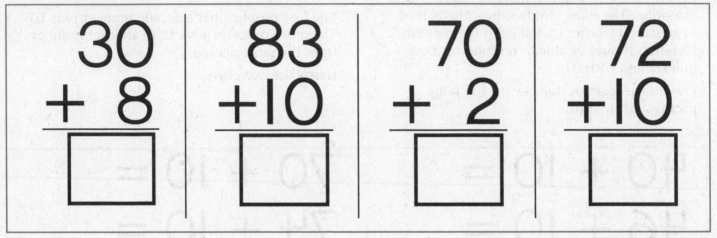

Lesson 105, Workbook

Teen Number Facts (Place-Value— Turn-Arounds—Plus 10)

As discussed in the Place Value track, the initial instruction on place value occurs with numbers 21–99, not with teens. On Lesson 102, the program introduces place value for teen numbers (10 + 6 = 16) and the turn-around equation (6 + 10 = 16). The turn-around equations add 10 to one-digit numbers (1 + 10 through 9 + 10).

Here's the first part of an exercise from Lesson 112:

EXERCISE 2: PLUS 10s TO ONE-PART NUMBERS

a. You've learned to say equations for teen numbers.
• Listen: 10 plus 6. What's 10 plus 6? (Signal.) *16.*
• So what's 6 plus 10? (Signal.) *16.*
• Yes, 6 plus 10 equals 16. Say the equation. (Tap 5.) *6 plus 10 equals 16.*
b. Listen: 10 plus 9. What's 10 plus 9? (Signal.) *19.*
• So what's 9 plus 10? (Signal.) *19.*
• Yes, 9 plus 10 equals 19. Say the equation. (Tap 5.) *9 plus 10 equals 19.*

---

from Lesson 112, Exercise 2

In step A, children answer the familiar place-value question "What's 10 plus 6?" Then the teacher presents the turn-around question "So what's 6 plus 10?" The teacher models the equation and directs children to say it. Step B repeats the steps with 10 + 9.

Teaching Note: Children should be firm on place-value addition. If children are firm on place-value addition, these tasks will go smoothly and children will extend their understanding of addition and commutativity to a new set of problems.

As children learn place value for teens, they also learn the rule that any teen number plus 10 yields a 20s number.

Here's the exercise from Lesson 105:

Lesson 105, Exercise 5

ALGEBRA ADDITION—$4 + \square = 9$

The most important application of counting from a number to a number occurs when children solve algebra addition problems.

The procedure for solving these problems involves these steps:

1. Read the problem (4 plus how many equals 9)

2. Get the first number going (*Fouuur*)

3. Count to the number after the equals (*5, 6, 7, 8, 9*) and make a counter line under the box for each number you count

4. Count the lines and write the number in the box to complete the equation

$$4 + \boxed{5} = 9$$
$$| | | | |$$

This problem type is difficult because the lines under the box are counted two different ways—lines for the number that is plused (1, 2, 3, 4) and lines that are added to 4 (5, 6, 7, 8, 9). Children have learned all of the counting skills they need to perform algebra addition problems before these problems are introduced.

The most common error children make applying the problem-solving strategy for algebra is to make a line for the number they start with, 4 in the example. When children do this, they end up with one too many lines and the wrong answer. Counting tasks that precede the introduction of algebra addition problems buttress against children making that mistake.

Here's an initial Algebra Addition exercise from Lesson 65:

---

a. (Write on the board:) [65:5A]

$$4 + \square = 9 \qquad 6 + \square = 10 \qquad 14 + \square = 17$$

You're going to read these problems. Then I'll show you how you work them.

- (Point to **4**.) Read the problem. Get ready. (Touch symbols.) *4 plus how many equals 9.*
- (Point to **6**.) Read the problem. Get ready. (Touch symbols.) *6 plus how many equals 10.*
- (Point to **14**.) The next problem has two-part numbers that start with 1. Read the problem. Get ready. (Touch symbols.) *14 plus how many equals 17.*
  (Repeat until firm.)

b. (Point to **4**.) This problem tells you to start with 4 (touch) and count to 9. (Touch.)
- (Point to **4**.) What does this problem tell you to start with? (Touch.) *4.*
- (Point to **9**.) What does it tell you to count to? (Touch.) *9.*

c. (Point to **6**.) This problem tells you to start with 6 (touch) and count to 10. (Touch.)
- (Point to **6**.) What does this problem tell you to start with? (Touch.) *6.*
- (Point to **10**.) What does it tell you to count to? (Touch.) *10.*

d. (Point to **4**.) What does this problem tell you to start with? (Touch.) *4.*
- (Point to **9**.) What does it tell you to count to? (Touch.) *9.*
- (Point to **6**.) What does this problem tell you to start with? (Touch.) *6.*
- (Point to **10**.) What does it tell you to count to? (Touch.) *10.*
- (Point to **14**.) What does this problem tell you to start with? (Touch.) *14.*
- (Point to **17**.) What does it tell you to count to? (Touch.) *17.*
  (Repeat step d until firm.)

e. (Point to **4**.) This problem tells you to start with 4 and count to 9. To work this problem, you make a line for each number you count. You'll count. I'll make the lines. Tell me to stop after I make the line for number 9.
- (Touch **4**.) Get it going. *Fouuur.* Count. (Make a line below the box after each number children count:) *5, 6, 7, 8, 9, stop.*
  (Teacher reference:) [65:5B1–5]

$$4 + \square = 9 \qquad 6 + \square = 10 \qquad 14 + \square = 17$$
$$|||||$$

f. (Point to **6**.) What does this problem tell you to start with? (Touch.) *6.*
- (Point to **10**.) What does it tell you to count to? (Touch.) *10.* You'll count. I'll make the lines. Tell me to stop after I make the line for number 10.
- (Touch **6**.) Get it going. *Siiix.* Count. (Make lines as children count.) *7, 8, 9, 10, stop.*
  (Teacher reference:) [65:5C1–4]

$$4 + \square = 9 \qquad 6 + \square = 10 \qquad 14 + \square = 17$$
$$||||| \qquad\qquad ||||$$

g. (Point to **14**.) What does this problem tell you to start with? (Touch.) *14.*
- (Point to **17**.) What does it tell you to count to? (Touch.) *17.*
- You'll count. I'll make the lines. Tell me to stop after I make the line for number 17.
- (Touch **14**.) Get it going. *Fourteeen.* Count. (Make lines as children count:) *15, 16, 17, stop.*
  (Teacher reference:) [65:5D1–3]

$$4 + \square = 9 \qquad 6 + \square = 10 \qquad 14 + \square = 17$$
$$||||| \qquad\qquad |||| \qquad\qquad |||$$

(Point to problems.) Remember how to make lines for this kind of plus problem.

In step A, children read the problems. In step B, the teacher models what the problem tells children to do. Then the teacher asks children the questions about what the problem tells them to do. Step D asks the questions for all three of the problems. In step E, the children get the first number going (4), count to the number after the equals (9), and say "Stop." The teacher makes a line under the box for each number the children count.

Teaching Note: Repeat step D until children are firm. Children must be firm on interpreting what the algebra addition problems tell children to do.

In steps E, F, and G, touch the first number for at least 1 second before making a line under the box. Children should get the number going when you touch it and keep it going as long as you touch it.

Make lines at the rate the children will be able to make lines when they work this kind of problem.

If children forget to say "Stop" when they are supposed to, tell them, "You've got to tell me when to stop." Point to the number after the equals and say, "After I make the lines for this number, remember to say *stop*." Erase the lines you made and repeat the task. Reinforce children when they remember to say "Stop."

After presenting the counting for all the examples, repeat the set if children had any problems.

If you are using the Board Displays, return to the frame at the beginning of a step to repeat the step.

By Lesson 73, children are working problems in which they make the lines and count the lines under the box to complete the equation.

Here's the first part of the exercise from Lesson 73:

EXERCISE 9: ALGEBRA ADDITION—*Solving Problems*

a. Touch the problem 16 plus how many equals 19 on worksheet 73. ✔
   (Teacher reference:)

   $16+\boxed{}=19$ $\quad$ $37+\boxed{}=42$

   You'll read each problem in that row, then work it.

   • Touch and read the first problem in that row. Get ready. (Tap 5.)
   16 plus how many equals 19.
   • What does the problem tell you to start with? (Signal.) *16.*
   • What does the problem tell you to count to? (Signal.) *19.*
   • Yes, the problem tells you to start with 16 and count to 19. What does the problem tell you to do? (Signal.) *Start with 16 and count to 19.*

b. You'll touch 16 and get it going. Then you'll count to 19 and make a line under the box for each number you count.

   • Pencils on 16. ✔
   • Get it going. *Sixteeen.* Count. (Tap 3.) (Children make a line for each number.) *17, 18, 19.*
   (Teacher reference:)

   $16+\underset{|||}{\boxed{}}=19$ $\quad$ $37+\boxed{}=42$

c. Count the lines under the box to yourself. Then write the number in the box.
   (Observe children and give feedback.)
   (Teacher reference:)

   $16+\underset{|||}{\boxed{3}}=19$ $\quad$ $37+\boxed{}=42$

   • Touch and read the equation. Get ready. (Tap 5.) *16 plus 3 equals 19.*
   • 16 plus how many equals 19? (Signal.) *3.*

---

from Lesson 73, Exercise 9

Teaching Note: In step A, children read the problem and indicate what the problem tells them to do—start with 16 and count to 19. Repeat this sequence several times. In step B, children count as they make lines under the box. In step C, children count the lines under the box to complete the equation. Then children read the equation and answer the question about the number they plused.

After children have learned algebra addition, they work sets of problems that present regular addition, subtraction, and algebra addition.

Presenting these mixed sets of problems is discussed in the next two tracks, **Take Away** and **Word Problems**.

Take Away (Subtraction)—Lessons 28–120

The track for Take Away (Subtraction) is less involved than the track for Plusing (Addition).

The first exercises that prepare children to work take-away problems are presented on Lesson 28. The track continues through the end of the program.

Here is the procedure children apply to solve subtraction problems:

1. Read the problem
2. Make counters for the number the problem starts with
3. Cross out the counters that are taken away
4. Count for the counters that are not crossed out
5. Write the number for those counters

Children first learn that they can take away lines by erasing them or by crossing them out.

Here's the first part of the introduction:

EXERCISE 4: TAKE AWAY

a. You've worked problems that plus. Some problems take away.
- (Write on the board:)   [28:4A]

|||||

- (Point to IIIII.) Here are 5 lines. How many lines are we starting with? (Touch.) *5.*
  We can take away lines by erasing them.
b. Count the lines I take away. Get ready. (Erase lines starting on far right.) *1, 2, 3.*
  (Teacher reference:)   [28:4B1–3]

||

- How many lines did I take away? (Signal.) *3.*
- We started with 5. Do we still have 5 lines? (Signal.) *No.*
c. (Point to II.) Here are the lines we ended up with.
  Raise your hand when you know how many lines we ended up with. ✔
- Everybody, how many did we end up with? (Touch.) *2.*
  Yes, we ended up with 2.
d. (Erase lines.) New problem: Count the lines as I make them. (Make 7 lines.) *1, 2, 3, 4, 5, 6, 7.*
  (Teacher reference:)   [28:4C1–7]

|||||||

- How many lines? (Signal.) *7.*
  Those are the lines we're starting with.

---

from Lesson 28, Exercise 4

e. I'm going to take away lines. Count the lines I take away. (Erase lines starting on far right.) *1, 2, 3, 4.*
  (Teacher reference:)   [28:4D1–4]

||||

- How many lines did I take away? (Signal.) *4.*
f. Now let's figure out how many lines we ended up with.
  I'll touch each line as you count them to **yourself**.
- Get ready. (Touch lines as children count to themselves.)
- Everybody, how many did we end up with? (Touch.) *3.*
  Yes, we ended up with 3.

from Lesson 28, Exercise 4

In steps A through C, the teacher presents 5 lines and erases 3 as children count the lines that are erased. Children then tell how many lines they end up with. In steps D through F, the teacher repeats the tasks in steps A through C with an example that starts with 7 lines, erases 4 lines and ends up with 3 lines.

Teaching Note: As discussed in the **Counting** track, children have counted different types of actions; these include clapping, making lines, dropping pennies, and dropping crayons. This exercise type requires children to count a new type of action—erasing lines.

This is also the first exercise in which children chain these different counting skills together to identify the number of lines the teacher started with, took away, and ended up with.

You'll write from left to right in steps A and D, but erase the lines from right to left (steps B and E). In later lessons children will cross out lines from right to left as per your model in this operation. If you are using the Board Displays, be aware that the sequence has one frame for each line that is erased or made. Step D's build-up has seven frames, one for each line that is made. Step E has four frames, one for each line that is erased.

On Lesson 31, children learn that they can take away lines by crossing them out. The goal is to teach them that a group of lines like this |||||╫ generates three numbers—the number the problem started with (6), the number the problem took away (2), and the number the problem ended up with (4). The group with some of the lines crossed out provides children with a map of how take away works.

Here's the exercise from Lesson 31:

EXERCISE 6: TAKE AWAY

a. (Write on the board:) [31:6A]

You've worked problems that take away by erasing lines. We can also take away by crossing out lines.

- (Point to IIIII.) Here are the lines we start with. What are these? (Signal.) *The lines we start with.*
  (Repeat until firm.)
  I'll take away 3 lines by crossing them out.
- How many lines will I take away? (Signal.) *3.*
  (Cross out from right to left as you count.) *1, 2, 3.*
  (Teacher reference:) [31:6B1–3]

- How many lines did I take away? (Signal.) *3.*
b. The lines we started with are all of the lines. Count them. Get ready. (Touch lines.) *1, 2, 3, 4, 5.*
- How many lines did we start with? (Signal.) *5.*
c. Then I took away some lines. Those are the lines that are crossed out.
- (Point to III.) Count the lines I took away. Get ready. (Touch crossed-out lines.) *1, 2, 3.*
- How many lines did I take away? (Signal.) *3.*
d. The lines that are not crossed out are the lines we ended up with.
- (Point to II.) Count these lines we ended up with. Get ready. (Touch lines not crossed out as children count.) *1, 2.*
- How many lines did we end up with? (Signal.) *2.*

e. I'll touch under lines. You tell me if they are the lines we took away or the lines we ended up with.
- (Run your finger below the crossed-out lines.) Are these the lines we took away or the lines we ended up with? (Signal.) *The lines we took away.*
- (Run your finger below the lines that are not crossed out.) Are these the lines we took away or the lines we ended up with? (Signal.) *The lines we ended up with.*
  (Repeat step e until firm.)

Lesson 31, Exercise 6

In step E, the teacher touches under the lines that are crossed out or those that are not crossed out and asks, "Are these the lines we took away or the lines we ended up with?"

Teaching Note: Children have learned that they plus lines by writing them from left to right. Taking away is the opposite of plusing. It follows that lines are taken away from right to left. Make the cross-out stroke slowly enough that children are able to count as each line is crossed out. If you are using the Board Displays, advance the three frames at a predictable pace.

Repeat step E, if necessary, and make it fun.

Once children learn the routine for groups of lines with some lines crossed out, they are able to look at the picture of the group and tell you what equation the picture represents—the number of lines originally in the group, the number of lines crossed out, and the number of lines the group ended up with. This is a profile of what take-away problems are.

On Lesson 37 children are presented with workbook tasks that involve groups with some lines crossed out.

$$IIII\text{H} =$$

Children complete the equation by writing the number of lines they end up with (3). This work prepares them for completing take-away equations of the form:

$$6 - 4 =$$
$$IIIIII$$

On the following lessons, children work similar problems in which they cross out a specified number of lines and complete an equation by showing the number of lines they end up with. Children also learn to read take-away directions. They read **– 4** as *take away 4* which tells them to cross out 4 lines.

The first written take-away problems are introduced on Lesson 42.

Here's part of the introduction:

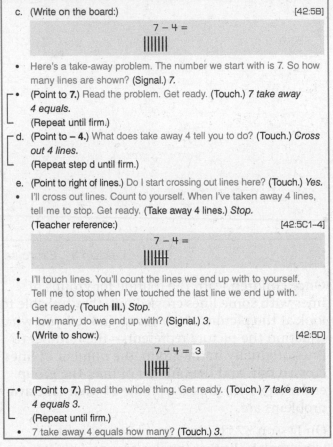

from Lesson 42, Exercise 5

In step C, children read the problem and identify the number of lines shown without counting them. In step D, children indicate what take away 4 tells them to do (cross out 4 lines).

The teacher crosses out lines in step E. Children count the lines the teacher takes away and tells the teacher when to stop. Then the teacher touches the lines that are not crossed out as children count to themselves. Children tell the teacher the number to complete the equation. In step F, children read the whole equation and answer the question, "7 take away 4 equals how many?"

Teaching Note: Children perform the same operations they performed earlier with lines. The only difference is that the symbols above the lines specify how many lines there are to start with and how many are taken away.

In this exercise, children are directed to count to themselves as the teacher crosses out lines. If children have trouble doing this, tell them they can count out loud, but they have to do it quietly.

On the following lessons, the teacher presents more take-away problems of the same form. Children also work problems that have lines, but no first number.

$$\square - 4 =$$
$$| | | | | |$$

Children count the lines the problem starts with and write the number in the box. They take away the lines and count the lines that are not crossed out to complete the equation. Children often have difficulty with the mechanics of making and crossing out lines. This problem type is mechanically similar to regular take-away problems (6 – 4 =), but it eliminates making lines and focuses on crossing out lines.

On Lesson 51, children work complete take-away problems in their Workbooks.

$$6 - 5 =$$

Children make lines for the number the problem starts with (6); cross out the number of lines the problem takes away (5); count the lines that the problem ends up with (1); and write the answer to complete the equation (1).

Connecting Math Concepts

PLUS/TAKE-AWAY DISCRIMINATION

Children continue to work groups of take-away problems through Lesson 55. On Lesson 56, they discriminate between problems that plus and take away.

Here's part of the exercise:

$$7 - 2 =$$

$$9 - 8 =$$

$$8 + 9 =$$

$$17 + 2 =$$

EXERCISE 4: PLUS/TAKE-AWAY DISCRIMINATION

Note: Do not display page until step b.

a. Some problems plus and some take away.
- For take-away problems, you make lines for the number you start with. Listen: For take-away problems, you make lines for which number? (Signal.) *The number you start with.* (Repeat until firm.)

b. (Display page and point to problems.)   [56:4A]
You'll read each problem and tell me which number you'd make lines for.
- (Point to 7 − 2 =.) Read the problem. Get ready. (Touch symbols.) *7 take away 2 equals.*
- Does that problem take away? (Touch.) *Yes.*
- So do you make lines for the number you start with? (Signal.) *Yes.*
- Listen: 7 take away 2 equals. Which number do you make lines for? (Signal.) *7.*

c. (Point to 8 + 9 =.) Read the problem. Get ready. (Touch symbols.) *8 plus 9 equals.*
- Does that problem take away? (Touch.) *No.*
- So do you make lines for the number you start with? (Signal.) *No.*
- Listen: 8 plus 9 equals. Which number do you make lines for? (Signal.) *9.*

from Lesson 56, Exercise 4

In step A, children answer the question "For take-away problems, you make lines for which number?"

In step B children read the first problem. Then they respond to the following set of three questions:

- Does that problem take away?
- So do you make lines for the number you start with?
- Listen: 7 take away 2. Which number do you make lines for?

Step C presents the tasks in step B for the next problem.

Teaching Note: For the remainder of the program children use the first two questions in steps B and C to determine how they solve problems. "Does that problem take away?" "So do you make lines for the number you start with?" If a problem takes away, children determine that they make lines for the number the problem starts with. If a problem does not take away, then children determine that they do not make lines for the number the problem starts out with, but for the number that is plused.

Make sure that children respond accurately and quickly to the sequence of four tasks you present for each problem. If children have trouble, model the answers for a sequence of four tasks, and then present those tasks to the children. Praise them for responding well.

Starting on Lesson 58, children work mixed sets of plus and take-away problems in their Workbooks until algebra addition problems are added to the mixed sets on Lesson 81.

Here's the set of problems children work on Lesson 81, followed by the Answer Key:

$$5 + \boxed{} = 9$$

$$7 - 6 = \boxed{}$$

$$26 + 3 = \boxed{}$$

$$8 + \boxed{} = 13$$

Lesson 81, Workbook

$$5 + \boxed{4} = 9$$
||||

$$7 - 6 = \boxed{1}$$
|卌

$$26 + 3 = \boxed{29}$$
|||

$$8 + \boxed{5} = 13$$
|||||

Lesson 81, Answer Key

TWO-DIGIT SUBTRACTION PROBLEMS

Two-digit subtraction problems are introduced on Lesson 101. Children follow the same procedure they use to work take-away problems with one-digit numbers, except they cross out Ts as well as lines.

Lesson 99 prepares children for this procedure with groups of Ts and lines that have some counters crossed out.

T T T T | | | | | | | +

Children simply count for the Ts and lines that are not crossed out (*Tennn, 11, 12, 13, 14, 15, 16, 17*).

Connecting Math Concepts

Subtraction with two-digit numbers begins on
Lesson 103:

EXERCISE 8: TAKING AWAY TWO-PART NUMBERS

a. Touch the problem 56 take away 52 equals on worksheet 103. ✔
(Teacher reference:)

$$56-52= \quad | \quad 44-14=$$

You're going to work take-away problems. The counters are already shown for the first problem. You're going to tell me how many Ts and lines you'll cross out to figure out the answer.

- Tell me the number you take away for the first problem. Get ready. (Signal.) *52.*
- How many Ts will you cross out? (Signal.) *5.*
- How many lines will you cross out? (Signal.) *2.*
- Put your pencil where you'll start crossing out the Ts. ✔
- Cross out 5 Ts.
  (Observe children and give feedback.)

b. Put your pencil where you'll start crossing out lines. ✔
- Cross out 2 lines.
  (Observe children and give feedback.)
  (Teacher reference:)

$$56-52= \quad | \quad 44-14=$$

c. Now you're going to count for the Ts and lines that are not crossed out.
- All of the Ts are crossed out. Touch and count for the lines. Get ready. (Tap 4.) *1, 2, 3, 4.*
  (Repeat until firm.)
- What's the answer? (Signal.) *4.*
- Complete the equation. ✔
  (Teacher reference:)

$$56-52=4 \quad | \quad 44-14=$$

- Touch and read the equation. Get ready. (Tap 5.) *56 take away 52 equals 4.*
  (Repeat until firm.)

d. Touch the problem 44 take away 14 equals.
You're going to work this take-away problem. First you'll make the counters. Then you'll cross out Ts and lines to figure out the answer.
- Tell me the number you'll make counters for. (Signal.) *44.*
- How many Ts will you make for 44? (Signal.) *4.*
- How many lines will you make for 44? (Signal.) *4.*
- Make the counters for the second problem. Put your pencil down when you're finished.
  (Observe children and give feedback.)
  (Teacher reference:)

$$56-52=4 \quad | \quad 44-14=$$

e. Look at the number you take away in the second problem. ✔
- What number do you take away? (Signal.) *14.*
- How many Ts will you cross out? (Signal.) *1.*
- How many lines will you cross out? (Signal.) *4.*
- Put your pencil where you'll start crossing out the T. ✔
- Cross out 1 T.
  (Observe children and give feedback.)

f. Put your pencil where you'll start crossing out lines. ✔
- Cross out 4 lines.
  (Observe children and give feedback.)
  (Teacher reference:)

$$56-52=4 \quad | \quad 44-14=$$

g. Now you're going to count for the Ts that are not crossed out.
- All of the lines are crossed out. Touch and count for the Ts. Get ready. (Tap.) *10, 20, 30.*
  (Repeat until firm.)
- What's the answer? (Signal.) *30.*
- Tell me the parts of 30. Get ready. (Signal.) *3 and zero.*
- Complete the equation. ✔
  (Answer key:)

$$56-52=4 \quad | \quad 44-14=30$$

- Touch and read the equation. Get ready. (Tap 5.) *44 take away 14 equals 30.*
  (Repeat until firm.)

Lesson 103, Exercise 8

The Ts and lines are already made for the first example. For the other problem, children make Ts and lines, then cross out some of them and figure out the answer to the problem.

Teaching Note: These are the same procedures used to solve single-digit problems. The "tricky" part is where the children put their pencil before crossing out the Ts that are taken away (step A). They need to put their pencil to the right of the last T (tens counter), not after the last ones counter.

In step D, children make counters for the first number for the problem 44 – 14 =. Make sure they respond correctly to the task "Tell me the number you'll make counters for."

COLUMN SUBTRACTION

After children have practiced working two-digit subtraction problems for a number of lessons, children work subtraction problems that are presented in columns, not rows. Children make the counters to the right of the number that is subtracted:

$$56 \; \text{T T T T T | | | | | |}$$
$$-\;15$$

Starting on Lesson 107 and continuing to the end of the program, children work a mix of two-digit addition and subtraction problems presented in columns and rows with very little teacher guidance.

Here's the set of problems from Lesson 115, followed by the Answer Key:

```
   44                    5
  -12                  +26
 [    ]                [    ]

      36-30= [    ]
```

Lesson 115, Workbook

```
   44  TTT∓||#           5   TT|||||||
  -12                  +26
 [32]                 [31]

      36-30= [ 6 ]
         ∓∓∓
         ||||||
```

Lesson 115, Answer Key

Word Problems

Addition word problems are introduced on Lesson 25 and continue through the end of the program. Subtraction word problems are introduced on Lesson 52. After Lesson 49, children use the following procedure to work word problems:

1. Translate the word problem into symbols

2. Solve the equation for the symbols

3. Use the solution of the equation to answer the question asked in the word problem

The part of the procedure that children need the most instruction and practice performing is translating the problems into symbols. The other parts will be relatively easy if children have mastered the counting and operational skills taught in preceding lessons. Children have practiced solving addition and subtraction equations in isolation. Children will have also used the solution of problems to answer questions.

ADDITION WORD PROBLEMS

The word problems that are first presented in Lesson 25 tell about somebody making two groups of lines. For the initial problems, children don't translate the problems into symbols. Children simply make the lines. Making the lines is a natural bridge between what children have been doing (counting two groups of lines) and what they do to translate word problems into two groups of lines.

Here's the exercise from Lesson 25:

EXERCISE 3: WORD PROBLEMS

a. I'll tell you a story. Listen: The teacher made two groups of lines.
- What did the teacher make? (Signal.) *Two groups of lines.*
  Yes, the teacher made two groups of lines.
  There were 4 lines in the first group and 3 lines in the other group.
- Listen again: There were 4 lines in the first group.
  How many lines were in the first group? (Signal.) *4.*
- There were 3 lines in the other group.
  How many lines were in the other group? (Signal.) *3.*

b. Listen again. Then you'll tell me how to make the groups. There were 4 lines in the first group and 3 lines in the other group.
- How many lines do I make for the first group? (Signal.) *4.*
- I'll make the lines. Count and tell me when to stop. Get ready. (Make 4 lines.) *1, 2, 3, 4, stop.*
  (Teacher reference:)   [25:3A1–4]

c. Yes, there were 4 lines in the first group and 3 lines in the other group.
- We made the first group. How many lines do I make for the **other** group? (Signal.) *3.*
- I'll make the lines. Count and tell me when to stop. Get ready. (Make 3 lines in second group.) *1, 2, 3, stop.*
  (Teacher reference:)   [25:3B1–3]

|||| |||

d. Now we can figure out how many lines are in both groups.
- Tell me to touch line 4. (Signal.) *Touch line 4.*
- (Touch line 4.) Get it going. *Fouuur.* (Touch lines in second group.) *5, 6, 7.*
  (Repeat until firm.)
- How many lines are in both groups? (Signal.) *7.*
  So the teacher made 7 lines altogether.
  How many lines did the teacher make altogether? (Signal.) *7.*

Lesson 25, Exercise 3

In step A, children say the numbers for each group: 4 for the first group and 3 for the other group. In steps B and C, children count the lines for the groups and tell the teacher when to stop making lines. In step D, children figure out the number of lines for both groups by directing the teacher to touch line 4, children get 4 going, and count-on for the lines in the other group.

On Lesson 34, children tell the teacher how to write the symbols for a word problem that tells about making lines.

Here's part of the exercise from Lesson 34:

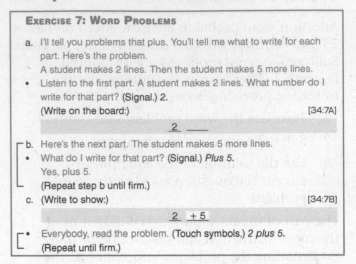

from Lesson 34, Exercise 7

In step A the teacher says the problem. Children identify the number for the first part.

In step B, children tell the teacher the symbols for "then the student makes 5 more lines." Children read the symbols for the problem in step C.

On Lesson 35, children figure out the answer to a word problem that tells about somebody making symbols. These problems are easy for children to conceptualize because they involve actions children have performed.

Here's part of the exercise:

d. I'll tell you another story. This time we'll work the whole problem. Listen: A student wrote 1 symbol. Then the student wrote 4 more symbols.
• Listen to the first part: A student wrote 1 symbol.
• What do I write for that part? (Signal.) *1.*
(Write on the board:) [35:9C]

$$\underline{1}\ \underline{}$$

e. Listen to the next part: Then the student wrote 4 more symbols.
• What do I write for that part? (Signal.) *Plus 4.*
(Write to show:) [35:9D]

$$\underline{1}\ \underline{+4}$$

f. We want to find out what 1 plus 4 **equals.** So what symbol do I write next? (Signal.) *Equals.*
(Repeat step f until firm.)
(Write to show:) [35:9E]

$$\underline{1}\ \underline{+4}\ \underline{=}$$

g. Now I make lines for one of the numbers. Which number? (Signal.) *4.*
• I'll make lines. Count to yourself and tell me when to stop. Get ready. (Make 4 lines.) *Stop.*
(Teacher reference:) [35:9F1–4]

$$\underline{1}\ \underline{+4}\ \underline{=}$$
$$||||$$

h. Now we'll count for both groups.
• (Touch 1.) Get it going. *Wuuun.* (Touch lines.) *2, 3, 4, 5.*
(Repeat until firm.)
• How many in both groups. (Signal.) *5.*
i. Where do I write 5? (Call on a child. Idea: *On the other side of the equals.*)
Yes, I write 5 on the other side of the equals.
• Everybody, where do I write 5? (Signal.) *On the other side of the equals.*
(Repeat until firm.)
(Write to show:) [35:9G]

$$\underline{1}\ \underline{+4}\ =\ \boxed{5}$$
$$||||$$

j. (Point to 1.) Everybody, read the whole thing. Get ready. (Touch symbols.) *1 plus 4 equals 5.*
• What does 1 plus 4 equal? (Signal.) *5.*
(Repeat until firm.)
So the student wrote 5 symbols. How many symbols did the student write? (Signal.) *5.*

from Lesson 35, Exercise 9

On later lessons, the scope of the problems progressively increases.

For example:

A teacher read 5 books. Then the teacher read 2 more books. (Lesson 41)

A man washed 6 windows. Then the man washed 6 more windows. (Lesson 43)

As the wording of the problems expands, children also learn to discriminate whether written problems accurately represent word problems. For instance the teacher reads two problems:

$$5 + 4 = \text{ and } 5 + 2 =$$

The teacher reads a word problem. Children touch the problem that has the right symbols. Here's the problem they listen to:

A teacher read 5 books. Then the teacher read 2 more books.

Children identify the symbols that represent the word problem and work it. The symbols for the problem are already written in the workbook, so children are able to practice translating the word problem into symbols without spending the considerable time writing symbols.

On Lesson 49 children write and work complete problems. Here's the first part of the exercise:

EXERCISE 10: WORD PROBLEMS—_Solving the Problem_

a. Touch the chair on your worksheet. ✔
   (Teacher reference:)

   I'll tell you plus word problems. You're going to write the symbols for the problems in the spaces next to the chair.
   Listen to the first problem: Mary wrote 6 letters. Then she wrote 3 more letters.
   Listen again: Mary wrote 6 letters. Then she wrote 3 more letters.
* Tell me all the symbols you'll write for that problem. Get ready. (Signal.) _6 plus 3 equals._
   Yes, 6 plus 3 equals.
* Again: Tell me the symbols you'll write for that problem. (Signal.) _6 plus 3 equals._
* Write 6 plus 3 equals in the space next to the chair.
   (Observe children and give feedback.)
b. Touch the number you'll make lines for. ✔
* How many lines will you make? (Signal.) _3._
* Touch the number you'll get going. ✔
* What number will you get going? (Signal.) _6._
c. Touch the number you'll make lines for again. ✔
* Make lines. Count for both groups and write the number to make the sides equal.
   (Observe children and give feedback.)
d. Touch and read the whole thing. Get ready. (Tap 5.) _6 plus 3 equals 9._
* Mary wrote 6 letters. Then she wrote 3 more letters. How many letters did Mary write in all? (Signal.) _9._
   (Teacher reference:)

   6+3=9
   III

---

from Lesson 49, Exercise 10

Teaching Note: For 20 lessons, children have written symbols for word problems or directed the teacher to write symbols. The most common mistake children may make when they write their first problem is to leave out the + or = sign. If they make mistakes, direct them to read the symbols they wrote and remind them, "Remember to make all the symbols for the problem."

By the end of the program, children are well practiced in solving action problems of the type above. They also do some work with problems that combine two groups. For example:

> Jan read 9 books. Vern read 4 books. How many books did the children read altogether?

SUBTRACTION WORD PROBLEMS

Word problems that involve subtraction begin on Lesson 52. Initially, children say the symbols for writing problems.

Here's part of the exercise from Lesson 52:

EXERCISE 4: TAKE-AWAY WORD PROBLEMS—
Writing Symbols for Problems

a. You're going to tell me how to write word problems that take away. These are not plus problems, but take-away problems.
• Will I write plus or take away for these problems? (Signal.) *Take away.*
b. Listen: There were 12 cookies on the table. Then the children took 9 of those cookies from the table. How many cookies ended up on the table?
• Listen to the first part: There were 12 cookies on the table. What do I write for that part? (Signal.) *12.*
(Write to show:)   [52:4A]

| 12 |

c. Listen to the next part: Then the children took 9 of those cookies from the table.
I write take away 9 for that part.
• Listen again: Then the children took 9 of those cookies from the table. What symbols do I write for that part? (Signal.) *Take away 9.*
(Repeat until firm.)
(Write to show:)   [52:4B]

| 12 – 9 |

d. Now I write equals. What do I write? (Signal.) *Equals.*
(Write to show:)   [52:4C]

| 12 – 9 = |

---

from Lesson 52, Exercise 4

In step B children listen to the whole problem then listen to the first part and indicate what to write:

> There were 12 cookies on the table.
> *12*

In step C, children indicate what to write for the next part of the problem:

> Then the children took 9 of those cookies from the table.
> *12 – 9*

In step D, children indicate that they write an equals:

> *12 – 9 =*

In the rest of the exercise, which is not shown here, the steps are repeated for two more take-away problems:

> There were 11 cookies on the table. Then a dog took 2 of those cookies from the table. *(11 – 2 =)*
> There were 13 coats in the closet. Then Mrs. Jones took 7 of those coats from the closet. *(13 – 7 =)*

These problems are relatively easy because they refer to taking away. If the dog took 2 of those cookies from the table, children name the symbols, "Take away 2".

On Lesson 57, children write symbols for a take-away problem and work it:

Lesson 57, Exercise 10

Then say the next part: "Listen: Then 2 of those fleas jumped off his back. What do you write for that part?" Although the problem does not have the words *take away*, the teacher has told the children that these are take-away problems, so children are prompted about the relationship of situations like fleas leaving the dog and taking away.

After children respond correctly to each part, say the whole problem and direct children to write the symbols for it as you say it, "Listen: A cat had 7 fleas on his back. Then 2 of those fleas jumped off his back. Say the symbols you'll write for that problem."

PLUS/TAKE-AWAY DISCRIMINATION

On Lesson 61, children discriminate between word problems that plus and word problems that take away. The teacher says problems and children identify whether each problem pluses or takes away. Here's part of the exercise:

---

from Lesson 61, Exercise 7

In each step, the teacher says the problem twice, then asks children "Do you plus or take away for that problem?"

In later lessons, children identify whether problems plus or take away, then identify the symbols they write to work the problem. As this work continues, children do Workbook activities in which they write symbols for word problems and work them. The problem sets include both plus and take-away problems. By Lesson 69, the directions that the teacher provides for working these problems are minimal.

Here's the exercise from Lesson 69:

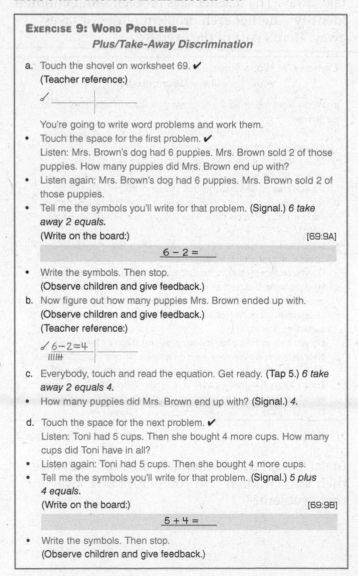

EXERCISE 9: WORD PROBLEMS—
Plus/Take-Away Discrimination

a. Touch the shovel on worksheet 69. ✔
    (Teacher reference:)

    ✓ _____|_____

    You're going to write word problems and work them.
•   Touch the space for the first problem. ✔
    Listen: Mrs. Brown's dog had 6 puppies. Mrs. Brown sold 2 of those puppies. How many puppies did Mrs. Brown end up with?
•   Listen again: Mrs. Brown's dog had 6 puppies. Mrs. Brown sold 2 of those puppies.
•   Tell me the symbols you'll write for that problem. (Signal.) *6 take away 2 equals.*
    (Write on the board:)   [69:9A]

    $$6 - 2 =$$

•   Write the symbols. Then stop.
    (Observe children and give feedback.)
b. Now figure out how many puppies Mrs. Brown ended up with.
    (Observe children and give feedback.)
    (Teacher reference:)

    ✓ 6−2=4 |
    |||||||

c. Everybody, touch and read the equation. Get ready. (Tap 5.) *6 take away 2 equals 4.*
•   How many puppies did Mrs. Brown end up with? (Signal.) *4.*

d. Touch the space for the next problem. ✔
    Listen: Toni had 5 cups. Then she bought 4 more cups. How many cups did Toni have in all?
•   Listen again: Toni had 5 cups. Then she bought 4 more cups.
•   Tell me the symbols you'll write for that problem. (Signal.) *5 plus 4 equals.*
    (Write on the board:)   [69:9B]

    $$5 + 4 =$$

•   Write the symbols. Then stop.
    (Observe children and give feedback.)

e. Now figure out how many cups Toni had in all.
    (Observe children and give feedback.)
    (Answer key:)

    ✓ 6−2=4 | 5+4=9
    ||||||| ||||

f. Everybody, read the equation. Get ready. (Tap 5.) *5 plus 4 equals 9.*
•   How many cups did Toni have in all? (Signal.) *9.*

Lesson 69, Exercise 9

In steps A through C, children work a take-away problem. In steps D through F, children work a plus problem.

TWO-DIGIT WORD PROBLEMS

Word problems that involve two-digit numbers begin on Lesson 108. Some are addition problems and some subtraction. The children write these problems in rows, although they know how to work them in columns as well.

If children have mastered the skills taught in earlier lessons, working word problems that involve two-digit numbers should be a relatively easy extension of what they can already do. Before Lesson 108, children have had a great deal of practice working two-digit addition and subtraction problems. They have also had a great deal of practice writing two-digit symbols. All of the other skills for working word problems that involve two-digit numbers are identical to the skills for working word problems presented in earlier lessons: Translate the word problem into symbols; complete the equation; use the solution of the equation to answer the question asked in the word problem.

Initially the program provides a great deal of structure to ensure that children write the correct symbols. This structure is needed because writing and working two-digit word problems consumes much more time than working word problems that involve one-digit numbers. The structure prevents children from getting the wrong answer by working a problem that was set up incorrectly.

As mentioned in the **Symbols** track, the word "digits" is not introduced in *CMC Level A*. The tens digit is referred to as "the first part" and the ones digit is referred to as "the other part."

Here's the workbook activity from Lesson 109:

b. Turn to the other side of worksheet 109 and touch the cup. ✔ You're going to write symbols for a word problem and work it.
- Touch the space for the problem. ✔
- Listen: There were 35 children playing in a park. 24 more children walked into the park. How many children ended up in the park?
- Listen again: There were 35 children playing in a park. 24 more children walked into the park. Tell me the symbols you'll write. Get ready. (Tap 4.) *35 plus 24 equals.*
- Tell me the parts of 35. Get ready. (Signal.) *3 and 5.*
- Tell me the parts of 24. Get ready. (Signal.) *2 and 4.*
- Write 35 plus 24 equals in the space for the problem. **(Observe children and give feedback.)**

c. Touch and read the symbols. Get ready. (Tap 4.) *35 plus 24 equals.*
- You'll make counters for one of the numbers. Which number? (Signal.) *24.*
- How many Ts? (Signal.) *2.*
- How many lines? (Signal.) *4.*
- Make 2 Ts and 4 lines below 24. Put your pencil down when you've done that much. **(Observe children and give feedback.)**

d. Touch the number you'll get going. Get it going and touch and count for the Ts and lines to yourself. Then write the answer. Put your pencil down when you've completed the equation. **(Observe children and give feedback.)** (Answer key:)

⊖ 35+24=59
  TTIIII

e. Touch and read the equation for the problem. Get ready. (Tap 5.) *35 plus 24 equals 59.*
- How many children ended up in the park? (Signal.) *59.*

from Lesson 109, Exercise 10

In step B, children say the numbers and the sign they'll write for the word problem and the digits (parts) of the two-digit numbers. In step C, children read the problem they wrote and answer questions about the counters they'll make. Then children make the counters. In step D, children work the problem and write the answer. In the last step, children read the equation and answer the question the problem asks.

Other Tracks

Some of the other tracks in *CMC Level A* teach Shapes (Lesson 85) and 3-D Objects (Lesson 107), Relative Position (Lesson 94) and Size (Lesson 98). Classification includes Ordering Objects (Lesson 112) and Comparing Objects (Lesson 118). Groups are Composed and Decomposed (Lesson 116). The Money track teaches four coins (pennies, nickels, dimes, quarters) and five bills ($1, $5, $10, $20, and $50). Bills are introduced beginning in Lesson 102. Coins, as seen earlier in Counting, Counting for Coins, are first introduced on Lesson 74.

Summary

As the Tracks section of this guide shows, *CMC Level A* takes children, including lower performers, far beyond what they are traditionally expected to be capable of learning. *CMC Level A* does this, not by placing unreasonable demands on what children are to learn or how quickly they are assumed to learn new discriminations and skills. Rather, *CMC Level A* designs the operations of addition, subtraction, and algebra addition so that any child with the skills needed to enter the program will learn these operations, including procedures for working 2-digit addition problems that traditionally require carrying, such as 38 + 43. The program hones the skills children need to identify the place value of the digits, to count sets that have both Ts (for tens) and lines (for ones), and to complete equations to solve problems.

The expectations for the children on any given lesson are simply to apply what has been taught earlier and to learn perhaps 10% new material. If these expectations are met, children will learn everything the program teaches, which will shape their understanding and intuitions about math far beyond any historical expectations.

Level A and Common Core State Standards

In Kindergarten, instructional time should focus on two critical areas: (1) representing and comparing whole numbers, initially with sets of objects; (2) describing shapes and space. More learning time in Kindergarten should be devoted to numbers than to other topics.

(1) Students use numbers, including written numerals, to represent quantities and to solve quantitative problems, such as counting objects in a set; counting out a given number of objects; comparing sets or numerals; and modeling simple joining and separating situations with sets of objects, or eventually with equations such as 5 + 2 = 7 and 7 – 2 = 5. (Kindergarten students should see addition and subtraction equations, and student writing of equations in kindergarten is encouraged, but it is not required.) Students choose, combine, and apply effective strategies for answering quantitative questions, including quickly recognizing the cardinalities of small sets of objects, counting and producing sets of given sizes, counting the number of objects in combined sets, or counting the number of objects that remain in a set after some are taken away.

(2) Students describe their physical world using geometric ideas (e.g., shape, orientation, spatial relations) and vocabulary. They identify, name, and describe basic two-dimensional shapes, such as squares, triangles, circles, rectangles, and hexagons, presented in a variety of ways (e.g., with different sizes and orientations), as well as three-dimensional shapes such as cubes, cones, cylinders, and spheres. They use basic shapes and spatial reasoning to model objects in their environment and to construct more complex shapes.

(Common Core State Standards Initiative, n.d., page 9)

Source: www.corestandards.org/assets/CCSSI_Math%20Standards.pdf

The *CMC Level A* and Common Core State Standards Chart (reproduced below), shows that *Level A* focuses on the primary critical area, numbers, but also addresses *all* the standards. This section of the guide focuses on standards that are **not covered** thoroughly in the preceding Tracks section.

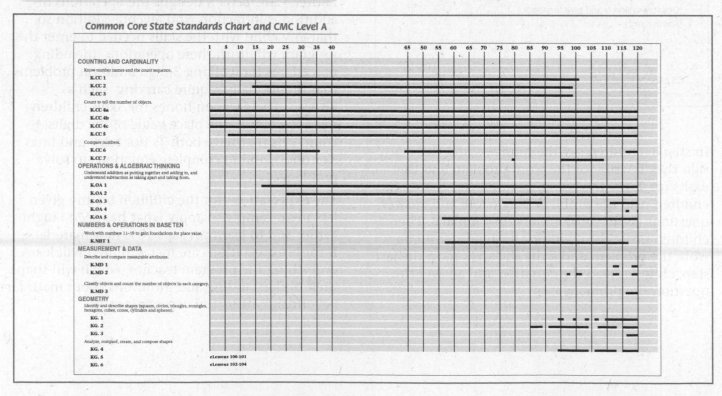

Counting and Cardinality (KCC)

Standards	Lessons
K.CC1	1–101
K.CC2	14–99
K.CC3	1–99
K.CC4a	1–120
K.CC4b	6–120
K.CC4c	1–120
K.CC5	6–120
K.CC6	19–36, 44–60, 116–120
K.CC7	79–80, 95–109

Exercises that address standards K.CC 4a and 4c begin on Lesson 1 and occur throughout the program. Standard 4c begins on Lesson 6. Below is more information how these standards are addressed in *CMC Level A*.

Common Core State Standards

Count to tell the number of objects.

K.CC4. Understand the relationship between numbers and quantities; connect counting to cardinality.

    a. When counting objects, say the number names in the standard order, pairing each object with one and only one number name and each number name with one and only one object.

    b. Understand that the last number name said tells the number of objects counted. The number of objects is the same regardless of their arrangement or the order in which they were counted.

    c. Understand that each successive number name refers to a quantity that is one larger.

(Common Core State Standards Initiative, n.d., page 11)

Source: www.corestandards.org/assets/CCSSI_Math%20Standards.pdf

Children first learn the relationship between numbers and quantities through counting exercises. Children learn to pair each object in a group with one and only one number name for groups of different types of objects. Children practice counting pictures of objects from right to left and from left to right. Children also count groups of real objects (coins and crayons) in various orders. Children demonstrate their understanding of the connection between counting and cardinality by counting. Children demonstrate their understanding that the number name for the last object in a group tells about the number of objects in the group and that each successive number name refers to a quantity that is one larger by correctly answering questions about the number of objects in the groups. A detailed discussion about how children learn these relationships and apply them to concrete objects, events and representations appears in this guide in the Tracks Counting section: **Reciting Counting Number**s (Rote Counting), **Counting Objects**, and **Counting Events**. Also see Equations and Equality, **Completing Equations** (page 78).

Common Core State Standards

Compare numbers.

K.CC6. Identify whether the number of objects in one group is greater than, less than, or equal to the number of objects in another group, e.g., by using matching and counting strategies.

By Lesson 115, children have learned extensive skills for counting and comparing numbers. Children have counted a variety of objects, counters and events. Children understand and have operated on the concepts of greater than and less than. Children have made more for numbers in problems that are "plused." Children have made less for numbers in problems that are "taken away."

In Lesson 116, children's skills are extended to identify whether the number of objects in one group is greater than, less than, or equal to the number of objects in another group. Children use both matching and counting strategies to perform these comparisons. In the first exercise, children compare the number of black cars to the number of blue cars and the number of girls to the number of boys. For each pair of groups, children

determine which is greater and which is less. On Lesson 117 children compare the number of cats to dogs, and the number of glasses to bottles.

On Lesson 118 children identify whether the number of objects in one group is greater than, less than, or equal to the number of objects in another group.

Here's the exercise from Lesson 118:

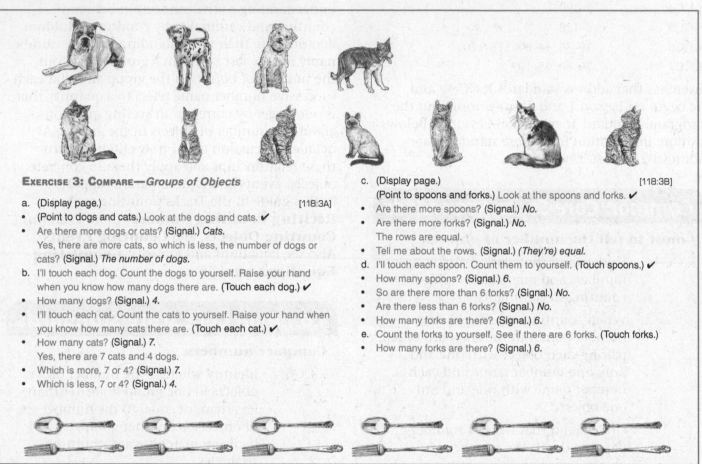

EXERCISE 3: COMPARE—*Groups of Objects*

a. (Display page.)   [118:3A]
- (Point to dogs and cats.) Look at the dogs and cats. ✔
- Are there more dogs or cats? (Signal.) *Cats.*
  Yes, there are more cats, so which is less, the number of dogs or cats? (Signal.) *The number of dogs.*
b. I'll touch each dog. Count the dogs to yourself. Raise your hand when you know how many dogs there are. (Touch each dog.) ✔
- How many dogs? (Signal.) *4.*
- I'll touch each cat. Count the cats to yourself. Raise your hand when you know how many cats there are. (Touch each cat.) ✔
- How many cats? (Signal.) *7.*
  Yes, there are 7 cats and 4 dogs.
- Which is more, 7 or 4? (Signal.) *7.*
- Which is less, 7 or 4? (Signal.) *4.*

c. (Display page.)   [118:3B]
  (Point to spoons and forks.) Look at the spoons and forks. ✔
- Are there more spoons? (Signal.) *No.*
- Are there more forks? (Signal.) *No.*
  The rows are equal.
- Tell me about the rows. (Signal.) *(They're) equal.*
d. I'll touch each spoon. Count them to yourself. (Touch spoons.) ✔
- How many spoons? (Signal.) *6.*
- So are there more than 6 forks? (Signal.) *No.*
- Are there less than 6 forks? (Signal.) *No.*
- How many forks are there? (Signal.) *6.*
e. Count the forks to yourself. See if there are 6 forks. (Touch forks.)
- How many forks are there? (Signal.) *6.*

Lesson 118, Exercise 3

Common Core State Standards

Compare numbers.

K.CC7.   Compare two numbers between 1 and 10 presented as written numerals.

Initially, the work children do to compare numbers is oral. On Lesson 79, children are asked questions of the type, "Which is more, 9 dogs or 8

dogs?" On lesson 81 children say pairs of numbers and identify which is more. Example:

The numbers are 9 and 10. Say those numbers. (S) *9 and 10.*
Which is more? (S) *10.*

Children have learned that the number that comes after another number when you count is the number that's more, so children should be able to perform these tasks easily.

On Lesson 83, children perform oral comparison tasks like the ones presented in Lessons 79–82. Then children compare pairs of written numbers by circling the number that is more. Only one of the written pairs in Lesson 83 contains numbers that are 10 or less.

Here's the exercise from Lesson 83:

EXERCISE 5: MORE

a. I'll say two numbers. You tell me which number is more.
- 6 and 9. Say the numbers. (Signal.) *6 and 9.*
- Which number is more? (Signal.) *9.*

b. 21 and 20. Say the numbers. (Signal.) *21 and 20.*
- Which number is more? (Signal.) *21.*
- 7 and 10. Say the numbers. (Signal.) *7 and 10.*
- Which number is more? (Signal.) *10.*
- 12 and 15. Say the numbers. (Signal.) *12 and 15.*
- Which number is more? (Signal.) *15.*
- 79 and 77. Say the numbers. (Signal.) *79 and 77.*
- Which number is more? (Signal.) *79.*
- (Repeat step b until firm.)

c. I'll say two weights. You'll say them. Then you'll tell me which is heavier.
- Listen: 77 pounds and 79 pounds. Say 77 pounds and 79 pounds. (Signal.) *77 pounds and 79 pounds.*
- Which is heavier? (Signal.) *79 pounds.*
- Listen: 10 pounds and 7 pounds. Say 10 pounds and 7 pounds. (Signal.) *10 pounds and 7 pounds.*
- Which is heavier? (Signal.) *10 pounds.*
- Listen: 6 ounces and 9 ounces. Say 6 ounces and 9 ounces. (Signal.) *6 ounces and 9 ounces.*
- Which is heavier? (Signal.) *9 ounces.*
- Listen: 15 ounces and 12 ounces. Say 15 ounces and 12 ounces. (Signal.) *15 ounces and 12 ounces.*
- Which is heavier? (Signal.) *15 ounces.*
- (Repeat step c until firm.)

d. Touch the hat on worksheet 83. ✔
(Teacher reference:)

  ✐ 6 9 | 21 20 | 79 77

Each problem shows two numbers. You're going to circle the number that is more.
- Touch the first numbers. ✔
You should be touching 6 and 9.
- Circle the number that is more.
(Observe children and give feedback.)
- Which number did you circle? (Signal.) *9.*
Yes, 9 is more than 6.

e. Touch the next two numbers. ✔
You should be touching 21 and 20.
- Circle the number that is more.
(Observe children and give feedback.)
- Which number did you circle? (Signal.) *21.*

f. Touch the last two numbers. ✔
You should be touching 79 and 77.
- Circle the number that is more.
(Observe children and give feedback.)
- Which number did you circle? (Signal.) *79.*

Lesson 83, Exercise 5

Comparing pairs of written numbers continues throughout the rest of the program. Children get sufficient practice comparing written numbers between 1 and 10 even though many of the other pairs that are presented contain larger 2-digit and sometimes 3-digit numbers.

Operations and Algebraic Thinking (K.OA)

Standards	Lessons
K.OA1	17–120
K.OA2	25–120
K.OA3	76–89, 111–120
K.OA4	116–117, 120
K.OA5	56–120

Exercises that address standards K.OA1 through 5 begin at Lesson 17, 25, 89, 116, and 56, respectively. Below are more details on K.OA3 and K.OA4 and how they are addressed in *CMC Level A*.

Common Core State Standards

Understand addition as putting together and adding to, and understand subtraction as taking apart and taking from.

K.OA3.   Decompose numbers less than or equal to 10 into pairs in more than one way.

(Common Core State Standards Initiative, n.d., page 11)

The strategy children learn to work subtraction problems involves decomposing a group of counters. Children make counters for the number the problem starts with. Children cross out counters for the number the problem subtracts. Children count the counters that are not crossed out to determine the answer. The lines children start with are decomposed into (a) the lines that are crossed out and (b) the lines that are not crossed out. Note that when children use the take-away strategy to work subtraction problems in *CMC Level A*, the result is both an equation and a drawing that represents the problem and the solution.

Here is an example:

$$7 - 3 = 4$$
||||‖‖‖

The picture shows 7 lines decomposed into 3 lines that are crossed out and 4 lines that are not crossed out, which depicts the equation.

A detailed discussion about how children learn to decompose numbers less than or equal to 10 appears in this guide in the Tracks section, **Take Away (Subtraction)**.

After children have mastered and applied this decomposition strategy for subtraction to hundreds of examples, children decompose numbers less than or equal to 10 into various pairs. What children infer from these exercises is how the number that's subtracted in each example is related to the answer. For these exercises, if the subtrahend increases by one, the answer decreases by one.

Here's the exercise from Lesson 116:

EXERCISE 6: DECOMPOSING GROUPS OF 7

a.   Touch the problem 7 take away 1 on worksheet 116. ✔
(Teacher reference:)

7–1=	7–2=
‖‖‖‖‖	‖‖‖‖‖
7–3=	7–4=
‖‖‖‖‖	‖‖‖‖‖

These problems are take-away problems. All of the problems start with 7, and the 7 lines are already shown for each problem.

• Work each problem and write the answer. Put your pencil down when you've completed all of the equations.
(Observe children and give feedback.)
(Answer key:)

7–1= 6	7–2= 5
‖‖‖‖‖	‖‖‖‖‖
7–3= 4	7–4= 3
‖‖‖‖‖	‖‖‖‖‖

b.   Check your work. You'll touch and read each equation.
• First equation. Get ready. (Tap 5.) *7 take away 1 equals 6.*
• Second equation. (Tap 5.) *7 take away 2 equals 5.*
• Third equation. (Tap 5.) *7 take away 3 equals 4.*
• Fourth equation. (Tap 5.) *7 take away 4 equals 3.*

c.   I'll say take-away problems that start with 7. Some of them are problems you just worked. See if you can tell me all of the answers.
• Listen: 7 take away zero. What's 7 take away zero? (Signal.) *7.*
• What's 7 take away 1? (Signal.) *6.*
• What's 7 take away 2? (Signal.) *5.*
• What's 7 take away 3? (Signal.) *4.*
• What's 7 take away 4? (Signal.) *3.*
(Repeat until firm.)

Lesson 116, Exercise 6

Here are details from the first cluster of Operations and Algebraic Thinking standards (K.OA4).

Children learn to plus any one-digit number to any other one-digit number using a counting strategy.

Here is an example:

$$4 + 5 = \boxed{}$$

For this problem children make the counters for the number that's plused.

$$4 + 5 = \boxed{}$$
$$\quad\ |\,|\,|\,|\,|$$

Children get the number the problem starts with going (4). Then they count-on for each line under 5 (5, 6, 7, 8, 9). The number the children end up with is the answer.

After children master applying the strategy for working plus problems, children learn how to solve algebra addition problems.

Here is an example:

$$4 + \boxed{} = 9$$
$$\quad\ \ |\,|\,|\,|\,|$$

Children get the number the problem starts with going (4). Then they count-on and make a line for each number they count (5 lines). Children stop making lines after they count to 9. Then children count the lines under the box and write the missing addend. A detailed discussion about how children learn this strategy and apply it appears in this guide in the Tracks section **Plus (Addition), Algebra Addition**.

After children master the algebra addition strategy and are able to discriminate when to use it, children are presented with a series of problems that start with numbers from 1 to 9, have a missing addend, and equal 10. What children infer from the exercises is how the missing addend in each example is related to each answer; if the starting number increases by one, the missing addend decreases by one.

Here's the exercise from Lesson 116:

EXERCISE 5: ALGEBRA ADDITION

a. Today's lesson is 116. What number? (Signal.) *116.*
- How many parts does 116 have? (Signal.) *3.*
- Think about the first part of **One hundred 16.** Then think of the parts for 16. Everybody, what are the parts for one hundred 16? (Signal.) *1, 1, and 6.*
b. The parts of 116 are 1, 1, and 6. Say the parts of 116 again. Get ready. (Signal.) *1, 1, and 6.*
- (Distribute unopened workbooks to children.) Open your workbook to worksheet 116. ✔
c. Touch the problem 6 plus how many equals 10. ✔
  All of these problems tell you to start at a number and count to 10. You'll work each problem; then you'll write the turn-around equation below it.
- Work the problems and write the turn-around equations. Put your pencil down when you've written all of the equations for this part.
  (Observe children and give feedback.)
  (Answer key:)

  $6+\boxed{4}=10$ $7+\boxed{3}=10$
  $\underset{|\,|\,|\,|}{}$
  $4+6=10$ $3+7=10$
  $8+\boxed{2}=10$ $9+\boxed{1}=10$
  $\underset{|\,|}{}$
  $2+8=10$ $1+9=10$

d. Check your work. You'll touch and read the equation for each problem. Then you'll touch and read the turn-around equation.
- First equation. Get ready. (Tap 5.) *6 plus 4 equals 10.*
- The turn-around equation. Get ready. (Tap 5.) *4 plus 6 equals 10.*
- Look at the first equation and listen. 6 plus how many equals 10? (Signal.) *4.*
- So 4 plus how many equals 10? (Signal.) *6.*
e. Read the second equation. Get ready. (Tap 5.) *7 plus 3 equals 10.*
- The turn-around equation. Get ready. (Tap 5.) *3 plus 7 equals 10.*
- Listen. 7 plus how many equals 10? (Signal.) *3.*
- So 3 plus how many equals 10? (Signal.) *7.*
f. Read the third equation. Get ready. (Tap 5.) *8 plus 2 equals 10.*
- The turn-around equation. Get ready. (Tap 5.) *2 plus 8 equals 10.*
- Listen. 8 plus how many equals 10? (Signal.) *2.*
- So 2 plus how many equals 10? (Signal.) *8.*
g. Read the fourth equation. Get ready. (Tap 5.) *9 plus 1 equals 10.*
- The turn-around equation. Get ready. (Tap 5.) *1 plus 9 equals 10.*
- Listen. 9 plus how many equals 10? (Signal.) *1.*
- So 1 plus how many equals 10? (Signal.) *9.*

Number and Operations in Base Ten (K.NBT)

Standard	Lessons
K.NBT 1	57–117

In Kindergarten Mathematics, there is a single standard for Number Operations in Base Ten: K.NBT1. *CMC Level A* begins to address this standard on Lesson 57.

Common Core State Standards

Work with numbers 11–19 to gain foundations for place value.

K.NBT1. Compose and decompose numbers from 11 to 19 into ten ones and some further ones ... understand that these numbers are composed of ten ones and one, two, three, four, five, six, seven, eight, or nine ones.

(Common Core State Standards Initiative, n.d., page 12)

In *CMC Level A*, children learn relationships of ones and tens that apply to all two-digit numbers. Children learn to make counters for two-digit numbers.

Examples:

25   14

Children also learn to say addition equations for each number. For 25, children say 20 plus 5 equals 25. For 14, children say 10 plus 4 equals 14. Children also learn to make counters for each number. Children make Ts for tens, and they make lines for the ones digit:

25   14
TTIIIII   TIIII

Children also learn to write numbers from a group of counters.

Examples:

Children count by tens for each T and count on by ones for each line. They write the number they end up with:

25   14
TTIIIII   TIIII

A detailed discussion about how children learn these relationships appears in this guide in the Tracks section **Place Value**. The discussion that deals with teen numbers appears in the Tracks Plus section, **Addition Facts, Teen-Number Facts** (page 102). The discussion that deals with making counters for numbers appears in the Tracks Counting section, **Counting Events**. The discussion that deals with writing numbers for counters appears in the Tracks Equality and Equations section, **Completing Equations** (page 78).

Children begin using their knowledge of place value to compose numbers from 11 to 19 into 10 ones and some further ones on Lesson 116. Children figure out the tens number represented by a T and lines. Children then decompose the T into 10 lines. Children demonstrate their understanding that the numbers in these examples are composed of 10 ones and 1, 2, 3, 4, 5, 6, 7, 8, or 9 ones by accurately identifying the teen number represented by the T and lines, and making the equivalent number of lines to complete the equation.

Here's the introductory exercise from Lesson 116:

EXERCISE 10: PLACE VALUE—*10 Ones for 1 Ten*

a. (Write on the board:)   [116:10A]

> TIII =

Here's a T and some lines. Count just the lines to yourself. Raise your hand when you know how many lines.
- How many lines? (Signal.) *3.*
  (Point to **T**.) You'll count for the T and the lines.
- Count for the T. (Touch.) *Tennn.* Count for the lines. (Touch.) *11, 12, 13.*

b. I'm going to write 13 as 13 lines.
- Count and tell me when to stop. (Make lines.) *1, 2, 3, 4, 5, 6, 7, 8, 9, 10, 11, 12, 13, stop.*
  (Teacher reference:)   [116:10B1–13]

> TIII = IIIIIIIIIIIII

- Are the sides equal? (Signal.) *Yes.*
- (Point to **TIII**.) How many are on this side? (Touch.) *13.*
- (Point to **IIIIIIIIIIIII**.) How many are on this side? (Touch.) *13.*

c. Touch the first problem with a T and lines on worksheet 116. ✔
  (Teacher reference:)

TIII = |TIIIII =

You're going to make lines for equations that have a T and some lines.
- Count for the T and the lines to yourself. Raise your hand when you know how many are on that side.
  (Observe children and give feedback.)
- How many? (Signal.) *15.*
- Make 15 lines on the other side of the equals.
  (Observe children and give feedback.)

d. Touch the next problem. ✔
- Count for the T and the lines to yourself. Raise your hand when you know how many are on that side.
- How many? (Signal.) *16.*
- Make 16 lines on the other side of the equals.
  (Observe children and give feedback.)
  (Answer key:)

TIII = IIIIIIIIIIIIIII | TIIIII = IIIIIIIIIIIIIIII

Lesson 116, Exercise 10

Measurement and Data (K.MD)

Standards	Lessons
K.MD1	112–114, 118–120
K.MD2	97–98, 100–102, 112–114, 118–120
K.MD3	116–120

Exercises that address the three Measurement and Data standards appear toward the end of the program. Here are more details on these three standards and how they are addressed in *CMC Level A*.

K.MD1 and 2 form a cluster.

Common Core State Standards

Describe and compare measurable attributes.

K.MD1. Describe measurable attributes of objects, such as length or weight. Describe several attributes of a single object.

K.MD2. Directly compare two objects with a measurable attribute in common to see which object has "more of" / "less of" the attribute and describe the difference.

(Common Core State Standards Initiative, n.d., page 12)

Children describe measurable attributes such as height, weight, age, and cost. Children compare different people's height and determine who is the tallest and who is the shortest. Children compare different people's weight and determine who is the heaviest and who is the lightest. In some exercises children describe three measurable attributes for individuals.

Here is the exercise from Lesson 119 that directs children to describe measurable attributes:

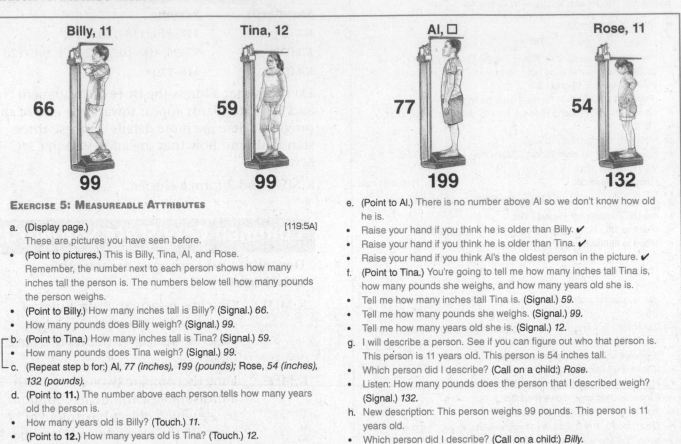

Billy, 11   66   99

Tina, 12   59   99

Al, □   77   199

Rose, 11   54   132

EXERCISE 5: MEASUREABLE ATTRIBUTES

a. (Display page.)   [119:5A]
   These are pictures you have seen before.
• (Point to pictures.) This is Billy, Tina, Al, and Rose. Remember, the number next to each person shows how many inches tall the person is. The numbers below tell how many pounds the person weighs.
• (Point to Billy.) How many inches tall is Billy? (Signal.) *66.*
• How many pounds does Billy weigh? (Signal.) *99.*
b. (Point to Tina.) How many inches tall is Tina? (Signal.) *59.*
• How many pounds does Tina weigh? (Signal.) *99.*
c. (Repeat step b for:) Al, *77 (inches), 199 (pounds);* Rose, *54 (inches), 132 (pounds).*
d. (Point to 11.) The number above each person tells how many years old the person is.
• How many years old is Billy? (Touch.) *11.*
• (Point to 12.) How many years old is Tina? (Touch.) *12.*
• (Point to Rose.) How many years old is Rose? (Touch.) *11.*

e. (Point to Al.) There is no number above Al so we don't know how old he is.
• Raise your hand if you think he is older than Billy. ✔
• Raise your hand if you think he is older than Tina. ✔
• Raise your hand if you think Al's the oldest person in the picture. ✔
f. (Point to Tina.) You're going to tell me how many inches tall Tina is, how many pounds she weighs, and how many years old she is.
• Tell me how many inches tall Tina is. (Signal.) *59.*
• Tell me how many pounds she weighs. (Signal.) *99.*
• Tell me how many years old she is. (Signal.) *12.*
g. I will describe a person. See if you can figure out who that person is. This person is 11 years old. This person is 54 inches tall.
• Which person did I describe? (Call on a child:) *Rose.*
• Listen: How many pounds does the person that I described weigh? (Signal.) *132.*
h. New description: This person weighs 99 pounds. This person is 11 years old.
• Which person did I describe? (Call on a child:) *Billy.*
• Everybody, how many inches tall is the person I described? (Signal.) *66.*

Lesson 119, Exercise 5

Here are more details on Measurement and Data Standard 3 (K.MD3) and how it is addressed in *CMC Level A*:

Common Core State Standards

Classify objects and count the number of objects in each category.

K.MD3.   Classify objects into given categories; count the number of objects in each category and sort the category by count.

(Common Core State Standards Initiative, n.d., page 12)

Children learn a strategy for organizing and sorting objects into different categories and expressing that data in a table. In Lesson 117, children are presented with the row of letters: M K M L K L L M L.

Children count the appearances of each letter and figure out equations: K=2, L=4, M=3.

Then children complete the table to sort the letters by the number of occurrences.

	2	3	4
Letter	K	M	L

On the following lessons, children sort letters and sort shapes by count. Children also complete tables on their worksheet, leaving spaces in the table blank if the space doesn't correspond to data for the row.

Here's the exercise from Lesson 118 in which children sort objects by count and complete a table to express the data:

EXERCISE 9: CLASSIFY OBJECTS—*Sort by Count*

a. Turn to the other side of worksheet 118 and touch the tree. ✔

- There's a row of letters next to the tree. I'll read the letters. Touch each letter as I read it. (Children touch letters.) P, R, R, T, P, S, P, P, R, T, P.

b. Touch the equations you'll complete to show how many Ps. ✔

- Count the Ps in the row of letters. Then complete the equation to show how many Ps there are. Put your pencil down when you've completed the equation for P.
(Observe children and give feedback.)

- How many Ps are there? (Signal.) *5.*

c. Complete the rest of the equations to show the number of Rs, Ss and Ts. Put your pencil down when you've completed the equations for all of the letters.
(Observe children and give feedback.)
(Teacher reference:)

P=5 R=3 S=1 T=2

- Touch the equation for R. ✔
- How many Rs are there? (Signal.) *3.*
- How many Ss are there? (Signal.) *1.*
- How many Ts are there? (Signal.) *2.*

d. Touch the row in the table that has the numbers 1, 2, 3, 4, 5, and 6. ✔ You're going to write the letters that go in the bottom row.

- Touch 1 in the top row of the table. ✔
- Look at the equations. Raise your hand when you know what letter equals 1. ✔
- What letter equals 1? (Signal.) *S.*
Later, you'll write S below 1 in the table.

e. Touch 2 in the top row of the table. ✔
- Raise your hand when you know what letter equals 2. ✔
- What letter equals 2? (Signal.) *T.*
- Touch where you'll write T. ✔

- Look at the equations. Raise your hand when you know what letter equals 1. ✔
- What letter equals 1? (Signal.) *S.*
Later, you'll write S below 1 in the table.

e. Touch 2 in the top row of the table. ✔
- Raise your hand when you know what letter equals 2. ✔
- What letter equals 2? (Signal.) *T.*
- Touch where you'll write T. ✔

f. Touch 3 in the top row. ✔
- Raise your hand when you know what letter equals 3. ✔
- What letter equals 3? (Signal.) *R.*
- Touch where you'll write R. ✔

g. Touch 4 in the top row. ✔
- Look at the equations. Raise your hand when you know if any letter equals 4. ✔
- Does any letter equal 4? (Signal.) *No.*
So you won't write any letter below 4.

h. Touch 5 in the top row. ✔
- Raise your hand when you know if any letter equals 5. ✔
- What letter equals 5? (Signal.) *P.*
- Touch where you'll write P. ✔

i. Touch 6 in the top row. ✔
- Raise your hand when you know if any letter equals 6. ✔
- Does any letter equal 6? (Signal.) *No.*
So do you write anything below 6? (Signal.) *No.*

j. Write the letters in the bottom row of the table. Remember, don't write anything below numbers that are not in the equations. Put your pencil down when you've completed the table.
(Observe children and give feedback.)
(Answer key:)

P=5 R=3 S=1 T=2
| Letter | S | T | R | | P | |

k. Check your work.
- Touch the number 1 in the table. ✔
- What letter did you write below 1? (Signal.) *S.*
- What did you write below 2? (Signal.) *T.*
- What did you write below 3? (Signal.) *R.*
- What did you write below 4? (Signal.) *Nothing.*
- What did you write below 5? (Signal.) *P.*
- What did you write below 6? (Signal.) *Nothing.*

Lesson 118, Exercise 9

Geometry (K.G)

Standards	Lessons
K.G1	94–96, 99–100, 103–104, 106–107, 109–120
K.G2	85–89, 91–104, 107–113, 115–120
K.G3	116–120
K.G4	94–104, 106–113, 115–120
K.G5	eLessons 100-101
K.G6	eLessons 102-104

Exercises that address Geometry standards K.G1 through 4 begin at Lesson 94, 85, 107, and 96, respectively. Standards K.G5 and K.G6 are addressed in the eLessons for *CMC Level A*.

Common Core State Standards

Identify and describe shapes (squares, circles, triangles, rectangles, hexagons, cubes, cones, cylinders, and spheres).

K.G1. Describe objects in the environment using names of shapes, and describe the relative positions of these objects using terms such as *above*, *below*, *beside*, *in front of*, *behind*, and *next to*.

K.G2. Correctly name shapes regardless of their orientations or overall size.

K.G3. Identify shapes as two-dimensional (lying in a plane, "flat") or three dimensional ("solid").

Analyze, compare, create, and compose shapes.

K.G4. Analyze and compare two- and three-dimensional shapes, in different sizes and orientations, using informal language to describe their similarities, differences, parts, and other attributes.

(Common Core State Standards Initiative, n.d., page 12)

The following table shows the two-dimensional (2-D) and three-dimensional (3-D) shapes children learn, the lesson in which the shapes are introduced, and the attributes children identify and use to discriminate the shapes.

Shape	Lesson Introduced	Attributes
Circle	85 (Exercise 2)	Perfectly round
Rectangle	86 (Exercise 3)	A box (4 sides)
Triangle	88 (Exercise 2)	3 sides (closed figure)
Square	96 (Exercise 4)	A rectangle with sides that are the same size
Hexagon	103 (Exercise 1)	6 sides (closed figure)
Cube	107 (Exercise 5)	Each face is a square 6 faces
Cylinder	110 (Exercise 4)	A circle on each end Both circles the same size
Cone	118 (Exercise 2)	A circle on one end, and a point on the other
Sphere	119 (Exercise 1)	Perfectly round, but not flat like a circle

When each shape is introduced, children identify it, then identify pictures of the shape that are different sizes and in different orientations. The program teaches children attributes of the new shape that discriminate it from familiar shapes. After learning the attributes, children are presented with a group of pictures containing the new shape and other familiar shapes. Children use the attributes to distinguish between the shapes and identify them.

The day after squares are introduced, children distinguish between rectangles and shapes that are not rectangles. To distinguish between rectangles and squares they determine if the sides of the rectangles are the same size.

Here is the exercise from Lesson 97:

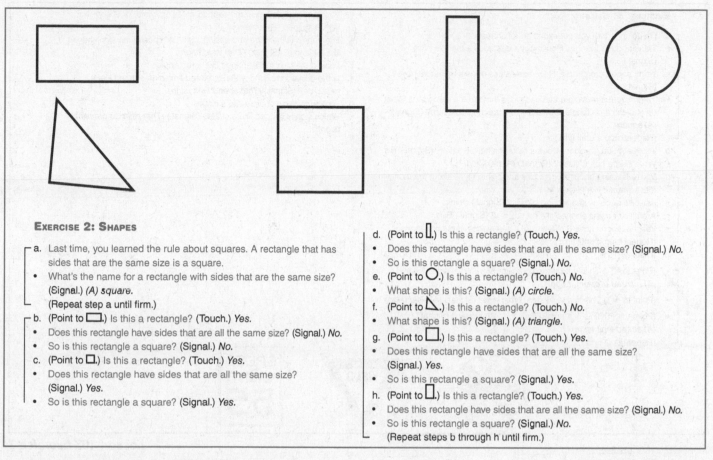

EXERCISE 2: SHAPES

a. Last time, you learned the rule about squares. A rectangle that has sides that are the same size is a square.
• What's the name for a rectangle with sides that are the same size? (Signal.) *(A) square.*
(Repeat step a until firm.)

b. (Point to ▭.) Is this a rectangle? (Touch.) *Yes.*
• Does this rectangle have sides that are all the same size? (Signal.) *No.*
• So is this rectangle a square? (Signal.) *No.*

c. (Point to ☐.) Is this a rectangle? (Touch.) *Yes.*
• Does this rectangle have sides that are all the same size? (Signal.) *Yes.*
• So is this rectangle a square? (Signal.) *Yes.*

d. (Point to ▯.) Is this a rectangle? (Touch.) *Yes.*
• Does this rectangle have sides that are all the same size? (Signal.) *No.*
• So is this rectangle a square? (Signal.) *No.*

e. (Point to ○.) Is this a rectangle? (Touch.) *No.*
• What shape is this? (Signal.) *(A) circle.*

f. (Point to △.) Is this a rectangle? (Touch.) *No.*
• What shape is this? (Signal.) *(A) triangle.*

g. (Point to ☐.) Is this a rectangle? (Touch.) *Yes.*
• Does this rectangle have sides that are all the same size? (Signal.) *Yes.*
• So is this rectangle a square? (Signal.) *Yes.*

h. (Point to ▯.) Is this a rectangle? (Touch.) *Yes.*
• Does this rectangle have sides that are all the same size? (Signal.) *No.*
• So is this rectangle a square? (Signal.) *No.*
(Repeat steps b through h until firm.)

Lesson 97, Exercise 2

Children begin applying their knowledge of shapes to describe objects in the environment on Lesson 107 (cubes: dice, ice cubes). In later lessons, children identify two-dimensional shapes from descriptions. Children also identify the shape of familiar two-dimensional objects, like bills, coins, and street signs and other two-dimensional objects in their environment.

Here's the exercise from Lesson 107:

EXERCISE 3: SHAPES

a. I'm going to ask you questions about shapes.
- Think about a triangle. How many sides does a triangle have? (Signal.) *3.*
- Think about a rectangle. How many sides does a rectangle have? (Signal.) *4.*
- Think about a rectangle that has sides that are the same size. What do you call a rectangle with sides that are the same size? (Signal.) *(A) square.*
  (Repeat step a until firm.)
b. Think about a hexagon. Does a hexagon have 6 sides? (Signal.) *Yes.*
- What shape has 6 sides? (Signal.) *(A) hexagon.*
c. You're learning about money: bills and coins. Think about the money that's shaped like circles. ✔
- Are bills or coins shaped like circles? (Signal.) *Coins.*
- Are bills or coins shaped like rectangles? (Signal.) *Bills.*
- Yes, hexagons have 6 sides. Say the sentence about hexagons.
  (Repeat step c until firm.)
d. (Display signs.)   [107:3A]
- (Point to ⊗.) This is a railroad crossing sign. What sign? (Signal.) *(A) railroad crossing (sign).*
- (Point to ▽.) This is a yield sign. What sign? (Signal.) *(A) yield (sign).*
- (Point to 55.) This is a speed limit sign. What sign? (Signal.) *(A) speed limit (sign).*
  (Repeat until firm.)

e. You're going to tell me if the railroad crossing, yield, or speed limit sign is shaped like a triangle.
- Everybody, is the railroad crossing, yield, or speed limit sign shaped like a triangle? (Signal.) *(The) yield (sign).*
f. Look at the sign that's shaped like a rectangle.
- Is the railroad crossing, yield, or speed limit sign shaped like a rectangle? (Signal.) *(The) speed limit (sign).*
g. Look at the sign shaped like a circle.
- Which sign is shaped like a circle? (Signal.) *(The) railroad crossing (sign).*

Lesson 107, Exercise 3

After children learn to identify 3-D shapes, they identify these shapes from descriptions and pictures. Children also identify objects in their environment that have these shapes.

EXERCISE 1: THREE DIMENSIONAL SHAPES— Cone and Sphere

a. (Display page.)   [120:1A]
  Some of these shapes are spheres.
- Say **sphere.** (Signal.) *Sphere.*
b. I'll touch the shapes. You'll say **sphere** or **not sphere** for each shape.
- (Point to small sphere.) Tell me. (Signal.) *Sphere.*
- (Repeat for:) Cylinder, *Not sphere;* Sphere, *Sphere;* Cone; *Not sphere.*
  (Repeat shapes that were not firm.)
c. (Point to cylinder.) Is this a sphere? (Touch.) *No.*
- How do you know? (Call on a child.) Ideas: *It has two ends; It's not round all over; The ends are circles.*

d. (Point to cone.) Is this a sphere? (Touch.) *No.*
- How do you know? (Call on a child.) Ideas: *It's not round on either end; It has a point.*
e. (Point to a sphere.) Is this a sphere? (Touch.) *Yes.*
- What is it? (Signal.) *(A) sphere.*
- How do you know it's a sphere? (Call on a child.) Idea: *It's the shape of a round ball.*
f. Raise your hand if you can name an object that's shaped like a sphere. (Call on a child. Ideas: *Ball, Earth, marble, etc.*)
g. Raise your hand if you can name an object that's shaped like a cylinder. (Call on a child. Ideas: *Pipe, can, barrel, etc.*)
h. Raise your hand if you can name an object that's shaped like a cone. (Call on a child. Ideas: *Ice cream cone, paper cone, clown hat, etc.*)
i. Raise your hand if you can name an object that's shaped like a cube. (Call on a child. Ideas: *Dice, ice cube, block, etc.*)

Lesson 120, Exercise 1

Children use their knowledge of 2-D and 3-D shapes to describe the relative position of shapes. Children are presented with a picture depicting 2-D and 3-D objects. Children identify the objects based on descriptions of their location, using words such as *below, above,* and *next to.* Children also describe the relative position of specific objects using these words.

Here's part of the exercise from Lesson 119:

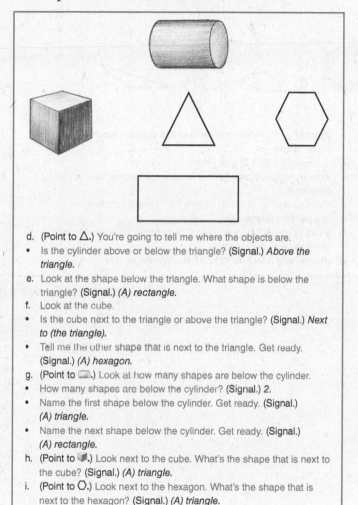

d. (Point to △.) You're going to tell me where the objects are.
• Is the cylinder above or below the triangle? (Signal.) *Above the triangle.*
e. Look at the shape below the triangle. What shape is below the triangle? (Signal.) *(A) rectangle.*
f. Look at the cube.
• Is the cube next to the triangle or above the triangle? (Signal.) *Next to (the triangle).*
• Tell me the other shape that is next to the triangle. Get ready. (Signal.) *(A) hexagon.*
g. (Point to ▭.) Look at how many shapes are below the cylinder.
• How many shapes are below the cylinder? (Signal.) *2.*
• Name the first shape below the cylinder. Get ready. (Signal.) *(A) triangle.*
• Name the next shape below the cylinder. Get ready. (Signal.) *(A) rectangle.*
h. (Point to ◼.) Look next to the cube. What's the shape that is next to the cube? (Signal.) *(A) triangle.*
i. (Point to ○.) Look next to the hexagon. What's the shape that is next to the hexagon? (Signal.) *(A) triangle.*

from Lesson 119, Exercise 3

Children demonstrate that they understand the difference between 2-D objects and 3-D objects in the first exercises that introduce 3-D objects. Children learn that a cube has a square on each face. Children identify the shape of the faces and identify the object that contains those faces. Children who can discriminate between faces and the whole object demonstrate that they understand the difference between 2-D shapes and 3-D shapes.

On the following lessons, children demonstrate this understanding with cubes, cylinders, and a variety of other objects that are minimally different from the objects they know, such as rectangular prisms and cones with the tip "cut off."

Children are also presented with pictures of 2-D and 3-D objects in perspective and determine if each object is flat or not flat.

Following is the exercise from Lesson 116 that presents tasks that isolate the discrimination between 2-D and 3-D objects:

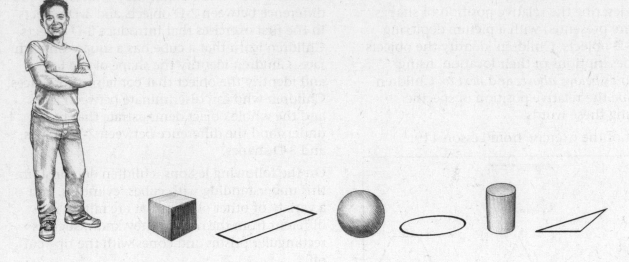

EXERCISE 4: TWO DIMENSIONS VERSUS THREE DIMENSIONS

a. (Display page and point to row.)   [116:4A]

Some of these shapes are flat, and the rest are not flat. If the man wanted to walk on top of the shapes that are not flat, he would have to step up. If the man wanted to walk on top of the flat objects, he wouldn't have to step up at all.

• (Point to cube.) What's the name of this shape? (Touch.) *(A) cube.*
• Is a cube flat? (Signal.) *No.*
• If the man walked on top of the cube, would he have to step up? (Signal.) *Yes.*

b. (Point to ⬭.) What's the name of this shape? (Touch.) *(A) rectangle.*
• Is a rectangle flat? (Signal.) *Yes.*
• If the man walked on top of the rectangle, would he have to step up? (Signal.) *No.*

c. (Point to ◯.) This is a ball. What is this? (Touch.) *(A) ball.*
• Is a ball flat? (Signal.) *No.*

d. (Point to ⬯.) What's the name of this shape? (Touch.) *(A) circle.*
• Is a circle flat? (Signal.) *Yes.*

e. (Repeat step d for:) *(A) cylinder, No; (A) triangle, Yes.*

Lesson 116, Exercise 4

Appendix A

Cumulative Test 1
Group-Administered
Group Summary Sheet
Page 1

Name

Total points for Part 1	21	21	21	21	21	21	21	21	21	21	21	21	21	21	21	21
Pass child if score is at least 18/21.	P N	P N	P N	P N	P N	P N	P N	P N	P N	P N	P N	P N	P N	P N	P N	P N
Total points for Part 2	9	9	9	9	9	9	9	9	9	9	9	9	9	9	9	9
Pass child if score is at least 6/9.	P N	P N	P N	P N	P N	P N	P N	P N	P N	P N	P N	P N	P N	P N	P N	P N
Total points for Part 3	9	9	9	9	9	9	9	9	9	9	9	9	9	9	9	9
Pass child if score is at least 6/9.	P N	P N	P N	P N	P N	P N	P N	P N	P N	P N	P N	P N	P N	P N	P N	P N
Total points for Part 4	12	12	12	12	12	12	12	12	12	12	12	12	12	12	12	12
Pass child if score is at least 8/12.	P N	P N	P N	P N	P N	P N	P N	P N	P N	P N	P N	P N	P N	P N	P N	P N
Total points for Part 5	10	10	10	10	10	10	10	10	10	10	10	10	10	10	10	10
Pass child if score is at least 5/10.	P N	P N	P N	P N	P N	P N	P N	P N	P N	P N	P N	P N	P N	P N	P N	P N
Total points for Part 6	12	12	12	12	12	12	12	12	12	12	12	12	12	12	12	12
Pass child if score is at least 8/12.	P N	P N	P N	P N	P N	P N	P N	P N	P N	P N	P N	P N	P N	P N	P N	P N
Total points for Part 7	14	14	14	14	14	14	14	14	14	14	14	14	14	14	14	14
Pass child if score is at least 7/14.	P N	P N	P N	P N	P N	P N	P N	P N	P N	P N	P N	P N	P N	P N	P N	P N
Total points for Part 8	8	8	8	8	8	8	8	8	8	8	8	8	8	8	8	8
Pass child if score is at least 8/8.	P N	P N	P N	P N	P N	P N	P N	P N	P N	P N	P N	P N	P N	P N	P N	P N
Total points for Part 9	20	20	20	20	20	20	20	20	20	20	20	20	20	20	20	20
Pass child if score is at least 16/20.	P N	P N	P N	P N	P N	P N	P N	P N	P N	P N	P N	P N	P N	P N	P N	P N
Total points for Part 10	10	10	10	10	10	10	10	10	10	10	10	10	10	10	10	10
Pass child if score is at least 5/10.	P N	P N	P N	P N	P N	P N	P N	P N	P N	P N	P N	P N	P N	P N	P N	P N
Total points for Part 11	14	14	14	14	14	14	14	14	14	14	14	14	14	14	14	14
Pass child if score is at least 7/14.	P N	P N	P N	P N	P N	P N	P N	P N	P N	P N	P N	P N	P N	P N	P N	P N
Total points for Part 12	20	20	20	20	20	20	20	20	20	20	20	20	20	20	20	20
Pass child if score is at least 15/20.	P N	P N	P N	P N	P N	P N	P N	P N	P N	P N	P N	P N	P N	P N	P N	P N

Name

Individually Administered Part 13–Counting																			
	Tasks	Points																	
	68–81	**7, 3** or **0**																	
	90–100	**7, 3** or **0**																	

Total points for **Part 13**	14	14	14	14	14	14	14	14	14	14	14	14	14	14	14	14
Pass child if score is at least 8/14.	P N	P N	P N	P N	P N	P N	P N	P N	P N	P N	P N	P N	P N	P N	P N	P N

Part 14–Symbol ID																			
Symbol	Teen #	What #																	
18	**3** or **0**	**3** or **0**																	
81	**3** or **0**	✕																	
12	**3** or **0**	**3** or **0**																	
21	**3** or **0**	✕																	
31	**3** or **0**	✕																	
13	**3** or **0**	**3** or **0**																	

Total points for **Part 14**	27	27	27	27	27	27	27	27	27	27	27	27	27	27	27	27
Pass child if score is at least 21/27.	P N	P N	P N	P N	P N	P N	P N	P N	P N	P N	P N	P N	P N	P N	P N	P N
Total number of points	200	200	200	200	200	200	200	200	200	200	200	200	200	200	200	200
Grade																

Name

Total points for **Part 1**	4	4	4	4	4	4	4	4	4	4	4	4	4	4	4	4
Pass child if score is at least 3/4.	P/N	P/N	P/N	P/N	P/N	P/N	P/N	P/N	P/N	P/N	P/N	P/N	P/N	P/N	P/N	P/N
Total points for **Part 2**	8	8	8	8	8	8	8	8	8	8	8	8	8	8	8	8
Pass child if score is at least 6/8.	P/N	P/N	P/N	P/N	P/N	P/N	P/N	P/N	P/N	P/N	P/N	P/N	P/N	P/N	P/N	P/N
Total points for **Part 3**	12	12	12	12	12	12	12	12	12	12	12	12	12	12	12	12
Pass child if score is at least 6/9.	P/N	P/N	P/N	P/N	P/N	P/N	P/N	P/N	P/N	P/N	P/N	P/N	P/N	P/N	P/N	P/N
Total points for **Part 4**	4	4	4	4	4	4	4	4	4	4	4	4	4	4	4	4
Pass child if score is at least 10/12.	P/N	P/N	P/N	P/N	P/N	P/N	P/N	P/N	P/N	P/N	P/N	P/N	P/N	P/N	P/N	P/N
Total points for **Part 5**	8	8	8	8	8	8	8	8	8	8	8	8	8	8	8	8
Pass child if score is at least 6/8.	P/N	P/N	P/N	P/N	P/N	P/N	P/N	P/N	P/N	P/N	P/N	P/N	P/N	P/N	P/N	P/N
Section A Subtotal	36	36	36	36	36	36	36	36	36	36	36	36	36	36	36	36
Total points for **Part 6**	6	6	6	6	6	6	6	6	6	6	6	6	6	6	6	6
Pass child if score is at least 5/6.	P/N	P/N	P/N	P/N	P/N	P/N	P/N	P/N	P/N	P/N	P/N	P/N	P/N	P/N	P/N	P/N
Total points for **Part 7**	10	10	10	10	10	10	10	10	10	10	10	10	10	10	10	10
Pass child if score is at least 8/10.	P/N	P/N	P/N	P/N	P/N	P/N	P/N	P/N	P/N	P/N	P/N	P/N	P/N	P/N	P/N	P/N
Total points for **Part 8**	5	5	5	5	5	5	5	5	5	5	5	5	5	5	5	5
Pass child if score is at least 4/5.	P/N	P/N	P/N	P/N	P/N	P/N	P/N	P/N	P/N	P/N	P/N	P/N	P/N	P/N	P/N	P/N
Total points for **Part 9**	6	6	6	6	6	6	6	6	6	6	6	6	6	6	6	6
Pass child if score is at least 4/6.	P/N	P/N	P/N	P/N	P/N	P/N	P/N	P/N	P/N	P/N	P/N	P/N	P/N	P/N	P/N	P/N
Total points for **Part 10**	12	12	12	12	12	12	12	12	12	12	12	12	12	12	12	12
Pass child if score is at least 10/12.	P/N	P/N	P/N	P/N	P/N	P/N	P/N	P/N	P/N	P/N	P/N	P/N	P/N	P/N	P/N	P/N
Total points for **Part 11**	4	4	4	4	4	4	4	4	4	4	4	4	4	4	4	4
Pass child if score is at least 3/4.	P/N	P/N	P/N	P/N	P/N	P/N	P/N	P/N	P/N	P/N	P/N	P/N	P/N	P/N	P/N	P/N
Total points for **Part 12**	4	4	4	4	4	4	4	4	4	4	4	4	4	4	4	4
Pass child if score is at least 2/4.	P/N	P/N	P/N	P/N	P/N	P/N	P/N	P/N	P/N	P/N	P/N	P/N	P/N	P/N	P/N	P/N
Total points for **Part 13**	8	8	8	8	8	8	8	8	8	8	8	8	8	8	8	8
Pass child if score is at least 6/7.	P/N	P/N	P/N	P/N	P/N	P/N	P/N	P/N	P/N	P/N	P/N	P/N	P/N	P/N	P/N	P/N
Total points for **Part 14**	8	8	8	8	8	8	8	8	8	8	8	8	8	8	8	8
Pass child if score is at least 6/8.	P/N	P/N	P/N	P/N	P/N	P/N	P/N	P/N	P/N	P/N	P/N	P/N	P/N	P/N	P/N	P/N
Section B Subtotal	63	63	63	63	63	63	63	63	63	63	63	63	63	63	63	63

Connecting Math Concepts

Name

Total points for **Part 15**	4	4	4	4	4	4	4	4	4	4	4	4	4	4	4
Pass child if score is at least 4/4.	P N	P N	P N	P N	P N	P N	P N	P N	P N	P N	P N	P N	P N	P N	P N
Total points for **Part 16**	8	8	8	8	8	8	8	8	8	8	8	8	8	8	8
Pass child if score is at least 6/8.	P N	P N	P N	P N	P N	P N	P N	P N	P N	P N	P N	P N	P N	P N	P N
Total points for **Part 17**	6	6	6	6	6	6	6	6	6	6	6	6	6	6	6
Pass child if score is at least 3/6.	P N	P N	P N	P N	P N	P N	P N	P N	P N	P N	P N	P N	P N	P N	P N
Total points for **Part 18**	4	4	4	4	4	4	4	4	4	4	4	4	4	4	4
Pass child if score is at least 3/4.	P N	P N	P N	P N	P N	P N	P N	P N	P N	P N	P N	P N	P N	P N	P N
Total points for **Part 19**	9	9	9	9	9	9	9	9	9	9	9	9	9	9	9
Pass child if score is at least 6/9.	P N	P N	P N	P N	P N	P N	P N	P N	P N	P N	P N	P N	P N	P N	P N
Total points for **Part 20**	7	7	7	7	7	7	7	7	7	7	7	7	7	7	7
Pass child if score is at least 7/7.	P N	P N	P N	P N	P N	P N	P N	P N	P N	P N	P N	P N	P N	P N	P N
Section C Subtotal	38	38	38	38	38	38	38	38	38	38	38	38	38	38	38
Total points for **Part 21**	4	4	4	4	4	4	4	4	4	4	4	4	4	4	4
Pass child if score is at least 2/4.	P N	P N	P N	P N	P N	P N	P N	P N	P N	P N	P N	P N	P N	P N	P N
Total points for **Part 22**	12	12	12	12	12	12	12	12	12	12	12	12	12	12	12
Pass child if score is at least 12/12.	P N	P N	P N	P N	P N	P N	P N	P N	P N	P N	P N	P N	P N	P N	P N
Total points for **Part 23**	8	8	8	8	8	8	8	8	8	8	8	8	8	8	8
Pass child if score is at least 6/7.	P N	P N	P N	P N	P N	P N	P N	P N	P N	P N	P N	P N	P N	P N	P N
Total points for **Part 24**	6	6	6	6	6	6	6	6	6	6	6	6	6	6	6
Pass child if score is at least 3/6.	P N	P N	P N	P N	P N	P N	P N	P N	P N	P N	P N	P N	P N	P N	P N
Total points for **Part 25**	8	8	8	8	8	8	8	8	8	8	8	8	8	8	8
Pass child if score is at least 4/8.	P N	P N	P N	P N	P N	P N	P N	P N	P N	P N	P N	P N	P N	P N	P N
Section D Subtotal	38	38	38	38	38	38	38	38	38	38	38	38	38	38	38

Name

Total points for **Part 26**	4	4	4	4	4	4	4	4	4	4	4	4	4	4	4	4
Pass child if score is at least 2/4.	P N	P N	P N	P N	P N	P N	P N	P N	P N	P N	P N	P N	P N	P N	P N	P N
Total points for **Part 27**	6	6	6	6	6	6	6	6	6	6	6	6	6	6	6	6
Pass child if score is at least 3/6.	P N	P N	P N	P N	P N	P N	P N	P N	P N	P N	P N	P N	P N	P N	P N	P N
Total points for **Part 28**	8	8	8	8	8	8	8	8	8	8	8	8	8	8	8	8
Pass child if score is at least 4/8.	P N	P N	P N	P N	P N	P N	P N	P N	P N	P N	P N	P N	P N	P N	P N	P N
Total points for **Part 29**	16	16	16	16	16	16	16	16	16	16	16	16	16	16	16	16
Pass child if score is at least 12/16.	P N	P N	P N	P N	P N	P N	P N	P N	P N	P N	P N	P N	P N	P N	P N	P N
Total points for **Part 30**	10	10	10	10	10	10	10	10	10	10	10	10	10	10	10	10
Pass child if score is at least 10/10.	P N	P N	P N	P N	P N	P N	P N	P N	P N	P N	P N	P N	P N	P N	P N	P N
Section E Subtotal	44	44	44	44	44	44	44	44	44	44	44	44	44	44	44	44
Total points for **Part 31**	12	12	12	12	12	12	12	12	12	12	12	12	12	12	12	12
Pass child if score is at least 12/12.	P N	P N	P N	P N	P N	P N	P N	P N	P N	P N	P N	P N	P N	P N	P N	P N
Total points for **Part 32**	16	16	16	16	16	16	16	16	16	16	16	16	16	16	16	16
Pass child if score is at least 12/16.	P N	P N	P N	P N	P N	P N	P N	P N	P N	P N	P N	P N	P N	P N	P N	P N
Section F Subtotal	28	28	28	28	28	28	28	28	28	28	28	28	28	28	28	28
Total points for **Part 33**	16	16	16	16	16	16	16	16	16	16	16	16	16	16	16	16
Pass child if score is at least 13/16.	P N	P N	P N	P N	P N	P N	P N	P N	P N	P N	P N	P N	P N	P N	P N	P N
Total points for **Part 34**	4	4	4	4	4	4	4	4	4	4	4	4	4	4	4	4
Pass child if score is at least 3/4.	P N	P N	P N	P N	P N	P N	P N	P N	P N	P N	P N	P N	P N	P N	P N	P N
Section G Subtotal	20	20	20	20	20	20	20	20	20	20	20	20	20	20	20	20
Total points for **Part 35**	12	12	12	12	12	12	12	12	12	12	12	12	12	12	12	12
Pass child if score is at least 10/12.	P N	P N	P N	P N	P N	P N	P N	P N	P N	P N	P N	P N	P N	P N	P N	P N
Total points for **Part 36**	11	11	11	11	11	11	11	11	11	11	11	11	11	11	11	11
Pass child if score is at least 9/11.	P N	P N	P N	P N	P N	P N	P N	P N	P N	P N	P N	P N	P N	P N	P N	P N
Total points for **Part 37**	10	10	10	10	10	10	10	10	10	10	10	10	10	10	10	10
Pass child if score is at least 8/10.	P N	P N	P N	P N	P N	P N	P N	P N	P N	P N	P N	P N	P N	P N	P N	P N
Section H Subtotal	33	33	33	33	33	33	33	33	33	33	33	33	33	33	33	33
Total number of points	300	300	300	300	300	300	300	300	300	300	300	300	300	300	300	300
Grade																

Appendix B

Mastery Test 1
Group Summary Sheet

Name

Group Remedy needed if > 1/4 children fail

Individually Administered

Part 1–Counting																		
	Item Count to 7	Points 16, 18, or o																
Total points for **Part 1**			16	16	16	16	16	16	16	16	16	16	16	16	16		
Retest child if < 8/16.			R N	R N	R N	R N	R N	R N	R N	R N	R N	R N	R N	R N	R N	R N	Y N

Part 2–Symbol Identification																		
	Item	Points																
	a. 4	6 or 0																
	b. 6	6 or 0																
	c. =	6 or 0																
	d. 2	6 or 0																
Total points for **Part 2**			24	24	24	24	24	24	24	24	24	24	24	24	24	24	
Retest child if < 24/24.			R N	R N	R N	R N	R N	R N	R N	R N	R N	R N	R N	R N	R N	R N	Y N

Part 3–Counting Objects																		
	Task	Points																
	a. Count shoes	16, 8 or 0																
	b. Answer questions	8 or 0																
Total points for **Part 3**			24	24	24	24	24	24	24	24	24	24	24	24	24	24	
Retest child if < 16/24.			R N	R N	R N	R N	R N	R N	R N	R N	R N	R N	R N	R N	R N	R N	Y N

Part 4–Next Number																		
	Task	Points																
	a. …4 Next #	6 or 0																
	b. …6 Next #	6 or 0																
	c. …3 Next #	6 or 0																
	d. …1 Next #	6 or 0																
	e. …5 Next #	6 or 0																
	f. …2 Next #	6 or 0																
Total points for **Part 4**			36	36	36	36	36	36	36	36	36	36	36	36	36	36	
Retest child if < 30/36.			R N	R N	R N	R N	R N	R N	R N	R N	R N	R N	R N	R N	R N	R N	Y N
Test 1 Total			100	100	100	100	100	100	100	100	100	100	100	100	100	100	
Grade																		

Mastery Test 2
Group Summary Sheet

Name →

Group Remedy needed if > 1/4 children fail →

Group-Administered

																Group Remedy
Total points for **Part 1** — Points **21, 14, 7,** or **0**	21	21	21	21	21	21	21	21	21	21	21	21	21	21	21	
Retest child if < 14/21.	R/N	R/N	R/N	R/N	R/N	R/N	R/N	R/N	R/N	R/N	R/N	R/N	R/N	R/N	R/N	Y/N

Individually Administered

Total points for **Part 2** — Points: **12, 9, 6, 3,** or **0**	12	12	12	12	12	12	12	12	12	12	12	12	12	12	12	
Retest child if < 12/12.	R/N	R/N	R/N	R/N	R/N	R/N	R/N	R/N	R/N	R/N	R/N	R/N	R/N	R/N	R/N	Y/N
Total points for **Part 3** — Points: **12, 9, 6, 3,** or **0**	12	12	12	12	12	12	12	12	12	12	12	12	12	12	12	
Retest child if < 9/12.	R/N	R/N	R/N	R/N	R/N	R/N	R/N	R/N	R/N	R/N	R/N	R/N	R/N	R/N	R/N	Y/N

Part 4–Symbol Identification

Item	Points																
5	4 or 0																
7	4 or 0																
3	4 or 0																
+	4 or 0																

Total Points for **Part 4**	16	16	16	16	16	16	16	16	16	16	16	16	16	16	16	
Retest child if < 16/16.	R/N	R/N	R/N	R/N	R/N	R/N	R/N	R/N	R/N	R/N	R/N	R/N	R/N	R/N	R/N	Y/N
Total points for **Part 5** — Points: **9, 4,** or **0**	9	9	9	9	9	9	9	9	9	9	9	9	9	9	9	
Retest child if < 4/9.	R/N	R/N	R/N	R/N	R/N	R/N	R/N	R/N	R/N	R/N	R/N	R/N	R/N	R/N	R/N	Y/N
Total points for **Part 6** — Points: **12, 8, 4,** or **0**	12	12	12	12	12	12	12	12	12	12	12	12	12	12	12	
Retest child if < 12/12.	R/N	R/N	R/N	R/N	R/N	R/N	R/N	R/N	R/N	R/N	R/N	R/N	R/N	R/N	R/N	Y/N
Total points for **Part 7** — Points per set: **0 to 9**	18	18	18	18	18	18	18	18	18	18	18	18	18	18	18	
Retest child if < 14/18.	R/N	R/N	R/N	R/N	R/N	R/N	R/N	R/N	R/N	R/N	R/N	R/N	R/N	R/N	R/N	Y/N
Test 2 Total	100	100	100	100	100	100	100	100	100	100	100	100	100	100	100	
Grade																

Mastery Test 3
Group Summary Sheet

Group Remedy needed if > 1/4 children fail

Group-Administered

Total points for **Part 1**	12	12	12	12	12	12	12	12	12	12	12	12	12	12	12	
Retest child if < 8/12.	R/N	R/N	R/N	R/N	R/N	R/N	R/N	R/N	R/N	R/N	R/N	R/N	R/N	R/N	R/N	Y/N

Total points for **Part 2**	12	12	12	12	12	12	12	12	12	12	12	12	12	12	12	
Retest child if < 12/12.	R/N	R/N	R/N	R/N	R/N	R/N	R/N	R/N	R/N	R/N	R/N	R/N	R/N	R/N	R/N	Y/N

Total points for **Part 3**	12	12	12	12	12	12	12	12	12	12	12	12	12	12	12	
Retest child if < 10/12.	R/N	R/N	R/N	R/N	R/N	R/N	R/N	R/N	R/N	R/N	R/N	R/N	R/N	R/N	R/N	Y/N

Total points for **Part 4**	6	6	6	6	6	6	6	6	6	6	6	6	6	6	6	
Retest child if < 6/6.	R/N	R/N	R/N	R/N	R/N	R/N	R/N	R/N	R/N	R/N	R/N	R/N	R/N	R/N	R/N	Y/N

Part 5—Count from 10 to 20

10 to 20	1st Try	2nd Try
Points	10	6

Total points for **Part 5**	10	10	10	10	10	10	10	10	10	10	10	10	10	10	10	
Retest child if < 6/10.	R/N	R/N	R/N	R/N	R/N	R/N	R/N	R/N	R/N	R/N	R/N	R/N	R/N	R/N	R/N	Y/N

Individually Administered
Part 6—Next #

#	Next #	Points
2	3	3 or 0
12	13	3 or 0
7	8	3 or 0
17	18	3 or 0
5	6	3 or 0
15	16	3 or 0
10	11	3 or 0

Total points for **Part 6**	21	21	21	21	21	21	21	21	21	21	21	21	21	21	21	
Retest child if < 18/21.	R/N	R/N	R/N	R/N	R/N	R/N	R/N	R/N	R/N	R/N	R/N	R/N	R/N	R/N	R/N	Y/N

Part 7—Symbol ID

	Symbol	Points
	1	3 or 0
	9	3 or 0
	0	3 or 0
	7 + 3	3 or 0
	6 + 2	3 or 0

Total points for **Part 7**	15	15	15	15	15	15	15	15	15	15	15	15	15	15	15	
Retest child if < 12/15.	R/N	R/N	R/N	R/N	R/N	R/N	R/N	R/N	R/N	R/N	R/N	R/N	R/N	R/N	R/N	Y/N

Part 8—Counting Both Groups

Item	Answer	Points
count	Seven, 8, 9, 10	4, 2, 0
how many	10	2 or 0
count	Six, 7, 8	4, 2, 0
how many	5	2 or 0

Total points for **Part 8**	12	12	12	12	12	12	12	12	12	12	12	12	12	12	12	
Retest child if < 10/12.	R/N	R/N	R/N	R/N	R/N	R/N	R/N	R/N	R/N	R/N	R/N	R/N	R/N	R/N	R/N	Y/N

Test 3 Total	100	100	100	100	100	100	100	100	100	100	100	100	100	100	100	
Grade																

Mastery Test 4
Group Summary Sheet

Name

Group Remedy needed if > 1/4 children fail

Group-Administered

Total points for **Part 1**	6	6	6	6	6	6	6	6	6	6	6	6	6	6	6	
Retest child if < 6/6.	R N	R N	R N	R N	R N	R N	R N	R N	R N	R N	R N	R N	R N	R N	R N	Y N
Total points for **Part 2**	10	10	10	10	10	10	10	10	10	10	10	10	10	10	10	
Retest child if < 8/10.	R N	R N	R N	R N	R N	R N	R N	R N	R N	R N	R N	R N	R N	R N	R N	Y N
Total points for **Part 3**	8	8	8	8	8	8	8	8	8	8	8	8	8	8	8	
Retest child if < 6/8.	R N	R N	R N	R N	R N	R N	R N	R N	R N	R N	R N	R N	R N	R N	R N	Y N
Total points for **Part 4**	5	5	5	5	5	5	5	5	5	5	5	5	5	5	5	
Retest child if < 5/5.	R N	R N	R N	R N	R N	R N	R N	R N	R N	R N	R N	R N	R N	R N	R N	Y N
Total points for **Part 5**	18	18	18	18	18	18	18	18	18	18	18	18	18	18	18	
Retest child if < 15/18.	R N	R N	R N	R N	R N	R N	R N	R N	R N	R N	R N	R N	R N	R N	R N	Y N
Total points for **Part 6**	9	9	9	9	9	9	9	9	9	9	9	9	9	9	9	
Retest child if < 9/9.	R N	R N	R N	R N	R N	R N	R N	R N	R N	R N	R N	R N	R N	R N	R N	Y N

Individually Administered

Part 7 Symbol ID	Symbol	Points																
	8	3 or 0																
	—	3 or 0																
	0	3 or 0																
	14	3 or 0																
	17	3 or 0																

| Total points for **Part 7** | 15 | 15 | 15 | 15 | 15 | 15 | 15 | 15 | 15 | 15 | 15 | 15 | 15 | 15 | 15 | |
| Retest child if < 15/15. | R N | R N | R N | R N | R N | R N | R N | R N | R N | R N | R N | R N | R N | R N | R N | Y N |

Part 8–Counting

Task	Points																	
	1st Try	2nd Try																
18 to 28	6	4 or 0																
28 to 38	6	4 or 0																

| Total points for **Part 8** | 12 | 12 | 12 | 12 | 12 | 12 | 12 | 12 | 12 | 12 | 12 | 12 | 12 | 12 | 12 | |
| Retest child if < 8/12. | R N | R N | R N | R N | R N | R N | R N | R N | R N | R N | R N | R N | R N | R N | R N | Y N |

Part 9 Next #	#	Answer	Points															
	4	5	2 or 0															
	9	10	2 or 0															
	19	20	2 or 0															
	29	30	2 or 0															
	25	26	2 or 0															
	13	14	2 or 0															

| Total points for **Part 9** | 12 | 12 | 12 | 12 | 12 | 12 | 12 | 12 | 12 | 12 | 12 | 12 | 12 | 12 | 12 | 12 |
| Retest child if < 10/12. | R N | R N | R N | R N | R N | R N | R N | R N | R N | R N | R N | R N | R N | R N | R N | Y N |

Part 10–Story Problem

Answer 6 + 3 =	Points 5 or 0																
Total points for **Part 10**	5	5	5	5	5	5	5	5	5	5	5	5	5	5	5	
Retest child if < 5/5.	R N	R N	R N	R N	R N	R N	R N	R N	R N	R N	R N	R N	R N	R N	R N	Y N

| **Test 4 Total** | 100 | 100 | 100 | 100 | 100 | 100 | 100 | 100 | 100 | 100 | 100 | 100 | 100 | 100 | 100 | |
| **Grade** | | | | | | | | | | | | | | | | |

Name

Group Remedy needed if > 1/4 children fail

	Name →															Group Remedy
Total points for **Part 1**	12	12	12	12	12	12	12	12	12	12	12	12	12	12	12	
Retest child if < 12/12.	R N	R N	R N	R N	R N	R N	R N	R N	R N	R N	R N	R N	R N	R N	R N	Y N
Total points for **Part 2**	12	12	12	12	12	12	12	12	12	12	12	12	12	12	12	
Retest child if < 10/12.	R N	R N	R N	R N	R N	R N	R N	R N	R N	R N	R N	R N	R N	R N	R N	Y N
Total points for **Part 3**	6	6	6	6	6	6	6	6	6	6	6	6	6	6	6	
Retest child if < 6/6.	R N	R N	R N	R N	R N	R N	R N	R N	R N	R N	R N	R N	R N	R N	R N	Y N
Total points for **Part 4**	12	12	12	12	12	12	12	12	12	12	12	12	12	12	12	
Retest child if < 9/12.	R N	R N	R N	R N	R N	R N	R N	R N	R N	R N	R N	R N	R N	R N	R N	Y N

Individually Administered Part 5–Symbol ID

Symbol	Teen #																
13	5 or 0																
11	5 or 0																
10	5 or 0																
12	5 or 0																

Total points for **Part 5**	20	20	20	20	20	20	20	20	20	20	20	20	20	20	20	
Retest child if < 15/20.	R N	R N	R N	R N	R N	R N	R N	R N	R N	R N	R N	R N	R N	R N	R N	Y N

Part 6–Counting

Tasks	Points																
37–47	7, 4, or 0																
47–57	7, 4, or 0																

Total points for **Part 6**	14	14	14	14	14	14	14	14	14	14	14	14	14	14	14	
Retest child if < 11/14.	R N	R N	R N	R N	R N	R N	R N	R N	R N	R N	R N	R N	R N	R N	R N	Y N

Part 7–Plus 1

Problem	Answer	Points																
9 + 1	10	4 or 0																
14 + 1	15	4 or 0																
17 + 1	18	4 or 0																
28 + 1	29	4 or 0																
13 + 1	14	4 or 0																
6 + 1	7	4 or 0																

Total points for **Part 7**	24	24	24	24	24	24	24	24	24	24	24	24	24	24	24	
Retest child if < 20/24.	R N	R N	R N	R N	R N	R N	R N	R N	R N	R N	R N	R N	R N	R N	R N	Y N
Test 5 Total	100	100	100	100	100	100	100	100	100	100	100	100	100	100	100	
Grade																

Connecting Math Concepts

Mastery Test 6
Group Summary Sheet

Group Remedy needed if > 1/4 children fail

Total points for **Part 1**	7	7	7	7	7	7	7	7	7	7	7	7	7	7	7	
Retest child if < 7/7.	R N	R N	R N	R N	R N	R N	R N	R N	R N	R N	R N	R N	R N	R N	R N	Y N
Total points for **Part 2**	18	18	18	18	18	18	18	18	18	18	18	18	18	18	18	
Retest child if < 15/18.	R N	R N	R N	R N	R N	R N	R N	R N	R N	R N	R N	R N	R N	R N	R N	Y N
Total points for **Part 3**	7	7	7	7	7	7	7	7	7	7	7	7	7	7	7	
Retest child if < 7/7.	R N	R N	R N	R N	R N	R N	R N	R N	R N	R N	R N	R N	R N	R N	R N	Y N
Total points for **Part 4**	10	10	10	10	10	10	10	10	12	10	10	10	10	10	10	
Retest child if < 8/10.	R N	R N	R N	R N	R N	R N	R N	R N	R N	R N	R N	R N	R N	R N	R N	Y N
Total points for **Part 5**	20	20	20	20	20	20	20	20	20	20	20	20	20	20	20	
Retest child if < 18/20.	R N	R N	R N	R N	R N	R N	R N	R N	R N	R N	R N	R N	R N	R N	R N	Y N

Individually Administered Part 6–Counting

Tasks	Points																
68–81	3,1 or 0																
90–100	3,1 or 0																

Total points for **Part 6**	6	6	6	6	6	6	6	6	6	6	6	6	6	6	6	
Retest child if < 4/6.	R N	R N	R N	R N	R N	R N	R N	R N	R N	R N	R N	R N	R N	R N	R N	Y N

Part 7–Counting by Tens

Task	Points																
10–100	6, 3, or 0																

Total points for **Part 7**	6	6	6	6	6	6	6	6	6	6	6	6	6	6	6	
Retest child if < 3/6.	R N	R N	R N	R N	R N	R N	R N	R N	R N	R N	R N	R N	R N	R N	R N	Y N

Part 8–Symbol ID

Symbol	Teen #	What #																
31	2 or 0	✕																
13	2 or 0	2 or 0																
16	2 or 0	2 or 0																
71	2 or 0	✕																
12	2 or 0	2 or 0																
15	2 or 0	2 or 0																

Total points for **Part 8**	20	20	20	20	20	20	20	20	20	20	20	20	20	20	20	
Retest child if < 18/20.	R N	R N	R N	R N	R N	R N	R N	R N	R N	R N	R N	R N	R N	R N	R N	Y N

Part 9–Clapping for Each Numeral

Numbers	Points																
24, 25, 26, 27	3, 2 or 0																
37, 38, 39, 40, 41	3, 2 or 0																

Total points for **Part 9**	6	6	6	6	6	6	6	6	6	6	6	6	6	6	6	
Retest child if < 4/6.	R N	R N	R N	R N	R N	R N	R N	R N	R N	R N	R N	R N	R N	R N	R N	Y N
Test 6 Total	100	100	100	100	100	100	100	100	100	100	100	100	100	100	100	
Grade																

Name

Group Remedy needed if > 1/4 children fail

Total points for **Part 1**	10	10	10	10	10	10	10	10	10	10	10	10	10	10	10	
Retest child if < 10/10.	R/N	R/N	R/N	R/N	R/N	R/N	R/N	R/N	R/N	R/N	R/N	R/N	R/N	R/N	R/N	Y/N
Total points for **Part 2**	12	12	12	12	12	12	12	12	12	12	12	12	12	12	12	
Retest child if < 12/12.	R/N	R/N	R/N	R/N	R/N	R/N	R/N	R/N	R/N	R/N	R/N	R/N	R/N	R/N	R/N	Y/N
Total points for **Part 3**	14	14	14	14	14	14	14	14	14	14	14	14	14	14	14	
Retest child if < 11/14.	R/N	R/N	R/N	R/N	R/N	R/N	R/N	R/N	R/N	R/N	R/N	R/N	R/N	R/N	R/N	Y/N
Total points for **Part 4**	16	16	16	16	16	16	16	16	16	16	16	16	16	16	16	
Retest child if < 12/16.	R/N	R/N	R/N	R/N	R/N	R/N	R/N	R/N	R/N	R/N	R/N	R/N	R/N	R/N	R/N	Y/N
Total points for **Part 5**	20	20	20	20	20	20	20	20	20	20	20	20	20	20	20	
Retest child if < 16/20.	R/N	R/N	R/N	R/N	R/N	R/N	R/N	R/N	R/N	R/N	R/N	R/N	R/N	R/N	R/N	Y/N

Individually Administered Part 6—Adding Tens

Task	Points
27–97	8,4 or 0

| Total points for **Part 6** | 8 | 8 | 8 | 8 | 8 | 8 | 8 | 8 | 8 | 8 | 8 | 8 | 8 | 8 | 8 | |
| Retest child if < 4/8. | R/N | R/N | R/N | R/N | R/N | R/N | R/N | R/N | R/N | R/N | R/N | R/N | R/N | R/N | R/N | Y/N |

Part 7—Tens Question

Symbol	Tasks	Points
	40 + 10	2 or 0
	47 + 10	3 or 0
	30 + 10	2 or 0
	37 + 10	3 or 0
	20 + 10	2 or 0
	27 + 10	3 or 0
	50 + 10	2 or 0
	57 + 10	3 or 0

Total points for **Part 7**	20	20	20	20	20	20	20	20	20	20	20	20	20	20	20	
Retest child if < 15/20.	R/N	R/N	R/N	R/N	R/N	R/N	R/N	R/N	R/N	R/N	R/N	R/N	R/N	R/N	R/N	Y/N
Test 7 Total	100	100	100	100	100	100	100	100	100	100	100	100	100	100	100	
Grade																

Mastery Test 8
Group Summary Sheet

Name → (student columns) ... Group Remedy needed if > 1/4 children fail

		Group Remedy
Total points for Part 1	20 (×15)	
Retest child if < 15/20.	R/N (×15)	Y/N
Total points for Part 2	6 (×15)	
Retest child if < 6/6.	R/N (×15)	Y/N
Total points for Part 3	9 (×15)	
Retest child if < 6/9.	R/N (×15)	Y/N
Total points for Part 4	15 (×15)	
Retest child if < 15/15.	R/N (×15)	Y/N
Total points for Part 5	12 (×15)	
Retest child if < 9/12.	R/N (×15)	Y/N

Individually Administered Part 6–Rote Counting

Task	Points
96–106	4, 3 or 0
120–130	4, 3 or 0

		Group Remedy
Total points for Part 6	8 (×15)	
Retest child if < 6/8.	R/N (×15)	Y/N

Part 7–Symbol ID

Symbol	Points
100	3 or 0
124	1, 2, 3 or 0
160	1, 2, 3 or 0
185	1, 2, 3 or 0
173	1, 2, 3 or 0

		Group Remedy
Total points for Part 7	15 (×15)	
Retest child if < 12/15.	R/N (×15)	Y/N

Part 8–Tens Question

Task	Points
4 + 3	3 or 0
1 + 16	3 or 0
0 + 8	3 or 0

		Group Remedy
Total points for Part 8	9 (×15)	
Retest child if < 3/9.	R/N (×15)	Y/N

Part 9–Adding

Task	Points
25, 35, 45, 55, 65	6, 4 or 0

		Group Remedy
Total points for Part 9	6 (×15)	
Retest child if < 4/6.	R/N (×15)	Y/N
Test 8 Total	100 (×15)	
Grade		

Mastery Test 9
Group Summary Sheet

Group Remedy needed if > 1/4 children fail

	1	2	3	4	5	6	7	8	9	10	11	12	13	14	15	Remedy
Total points for Part 1	20	20	20	20	20	20	20	20	20	20	20	20	20	20	20	
Retest child if < 20/20.	R/N	R/N	R/N	R/N	R/N	R/N	R/N	R/N	R/N	R/N	R/N	R/N	R/N	R/N	R/N	Y/N
Total points for Part 2	8	8	8	8	8	8	8	8	8	8	8	8	8	8	8	
Retest child if < 8/8.	R/N	R/N	R/N	R/N	R/N	R/N	R/N	R/N	R/N	R/N	R/N	R/N	R/N	R/N	R/N	Y/N
Total points for Part 3	6	6	6	6	6	6	6	6	6	6	6	6	6	6	6	
Retest child if < 6/6.	R/N	R/N	R/N	R/N	R/N	R/N	R/N	R/N	R/N	R/N	R/N	R/N	R/N	R/N	R/N	Y/N
Total points for Part 4	12	12	12	12	12	12	12	12	12	12	12	12	12	12	12	
Retest child if < 9/12.	R/N	R/N	R/N	R/N	R/N	R/N	R/N	R/N	R/N	R/N	R/N	R/N	R/N	R/N	R/N	Y/N
Total points for Part 5	16	16	16	16	16	16	16	16	16	16	16	16	16	16	16	
Retest child if < 12/16.	R/N	R/N	R/N	R/N	R/N	R/N	R/N	R/N	R/N	R/N	R/N	R/N	R/N	R/N	R/N	Y/N
Total points for Part 6	6	6	6	6	6	6	6	6	6	6	6	6	6	6	6	
Retest child if < 4/6.	R/N	R/N	R/N	R/N	R/N	R/N	R/N	R/N	R/N	R/N	R/N	R/N	R/N	R/N	R/N	Y/N

Individually Administered Part 7—Identifying Shapes

Shape	Points																
○	1 or 0																
▯	1 or 0																
▽	1 or 0																
▭	1 or 0																
○	1 or 0																
△	1 or 0																

	1	2	3	4	5	6	7	8	9	10	11	12	13	14	15	Remedy
Total points for Part 7	6	6	6	6	6	6	6	6	6	6	6	6	6	6	6	
Retest child if < 5/6.	R/N	R/N	R/N	R/N	R/N	R/N	R/N	R/N	R/N	R/N	R/N	R/N	R/N	R/N	R/N	Y/N

Part 8—Symbol ID

Symbol	Points																
186	2 or 0																
116	2 or 0																
141	2 or 0																
107	2 or 0																
150	2 or 0																
109	2 or 0																

	1	2	3	4	5	6	7	8	9	10	11	12	13	14	15	Remedy
Total points for Part 8	12	12	12	12	12	12	12	12	12	12	12	12	12	12	12	
Retest child if < 10/12.	R/N	R/N	R/N	R/N	R/N	R/N	R/N	R/N	R/N	R/N	R/N	R/N	R/N	R/N	R/N	Y/N

Part 9—Turn-Arounds

Symbol	Points																
3 + 20	7 or 2																
1 + 16	7 or 2																

	1	2	3	4	5	6	7	8	9	10	11	12	13	14	15	Remedy
Total points for Part 9	14	14	14	14	14	14	14	14	14	14	14	14	14	14	14	
Retest child if < 11/14.	R/N	R/N	R/N	R/N	R/N	R/N	R/N	R/N	R/N	R/N	R/N	R/N	R/N	R/N	R/N	Y/N
Test 9 Total	100	100	100	100	100	100	100	100	100	100	100	100	100	100	100	
Grade																

Connecting Math Concepts

Mastery Test 10
Group Summary Sheet

Name

Group Remedy needed if > 1/4 children fail

Total points for **Part 1**	14	14	14	14	14	14	14	14	14	14	14	14	14	14	14	
Retest child if < 11/14.	R N	R N	R N	R N	R N	R N	R N	R N	R N	R N	R N	R N	R N	R N	R N	Y N
Total points for **Part 2**	12	12	12	12	12	12	12	12	12	12	12	12	12	12	12	
Retest child if < 12/12.	R N	R N	R N	R N	R N	R N	R N	R N	R N	R N	R N	R N	R N	R N	R N	Y N
Total points for **Part 3**	9	9	9	9	9	9	9	9	9	9	9	9	9	9	9	
Retest child if < 9/9.	R N	R N	R N	R N	R N	R N	R N	R N	R N	R N	R N	R N	R N	R N	R N	Y N
Total points for **Part 4**	16	16	16	16	16	16	16	16	16	16	16	16	16	16	16	
Retest child if < 13/16.	R N	R N	R N	R N	R N	R N	R N	R N	R N	R N	R N	R N	R N	R N	R N	Y N
Total points for **Part 5**	10	10	10	10	10	10	10	10	10	10	10	10	10	10	10	
Retest child if < 8/10.	R N	R N	R N	R N	R N	R N	R N	R N	R N	R N	R N	R N	R N	R N	R N	Y N

Individually Administered Part 6–Counting Backward

	Task	Points																	
	10–1	8, 4 or 1																	

Total points for **Part 6**	8	8	8	8	8	8	8	8	8	8	8	8	8	8	8	
Retest child if < 4/8.	R N	R N	R N	R N	R N	R N	R N	R N	R N	R N	R N	R N	R N	R N	R N	Y N

Part 7–Next Number

	Task	Answer	Points																	
	6, 5	4	2 or 0																	
	8, 7	6	2 or 0																	
	4, 3	2	2 or 0																	
	7, 6	5	2 or 0																	
	3, 2	1	2 or 0																	

Total points for **Part 7**	10	10	10	10	10	10	10	10	10	10	10	10	10	10	10	
Retest child if < 8/10.	R N	R N	R N	R N	R N	R N	R N	R N	R N	R N	R N	R N	R N	R N	R N	Y N

Part 8–Symbol ID

Task	Answer	Points																	
second	girl	1 or 0																	
fifth	cat	1 or 0																	
third	horse	1 or 0																	
dog	fourth	2 or 0																	
boy	first	1 or 0																	

Total points for **Part 8**	5	5	5	5	5	5	5	5	5	5	5	5	5	5	5	
Retest child if < 4/5.	R N	R N	R N	R N	R N	R N	R N	R N	R N	R N	R N	R N	R N	R N	R N	Y N

Part 9–Shapes

	Is☐?	Shapes?	Is☐?	Points																	
△	No			2, 1, 0																	
☐	Yes		Yes	2, 1, 0																	
▭	Yes		No	2, 1, 0																	
○	No			2, 1, 0																	
△	No			2, 1, 0																	
○	Yes		Yes	2, 1, 0																	

Total points for **Part 9**	12	12	12	12	12	12	12	12	12	12	12	12	12	12	12	
Retest child if < 11 /12.	R N	R N	R N	R N	R N	R N	R N	R N	R N	R N	R N	R N	R N	R N	R N	Y N

Part 10–Turn-Arounds

Task	Answer	Points																	
What coin?	quarter	2 to 0																	
How much?	25 cents	2 to 0																	

Total points for **Part 10**	4	4	4	4	4	4	4	4	4	4	4	4	4	4	4	
Retest child if < 4/4.	R N	R N	R N	R N	R N	R N	R N	R N	R N	R N	R N	R N	R N	R N	R N	Y N
Test 10 Total	100	100	100	100	100	100	100	100	100	100	100	100	100	100	100	
Grade																

Mastery Test 11
Group Summary Sheet

Name ... *Group Remedy needed if > 1/4 children fail*

	Name (14 students)	Group Remedy
Total points for Part 1 — Retest child if < 8/10.	10 10 10 10 10 10 10 10 10 10 10 10 10 10 (R/N each)	Y/N
Total points for Part 2 — Retest child if < 10/10.	10 10 10 10 10 10 10 10 10 10 10 10 10 10 (R/N each)	Y/N
Total points for Part 3 — Retest child if < 8/10.	10 10 10 10 10 10 10 10 10 10 10 10 10 10 (R/N each)	Y/N
Total points for Part 4 — Retest child if < 8/8.	8 8 8 8 8 8 8 8 8 8 8 8 8 8 (R/N each)	Y/N
Total points for Part 5 — Retest child if < 8/10.	10 10 10 10 10 10 10 10 10 10 10 10 10 10 (R/N each)	Y/N
Total points for Part 6 — Retest child if < 18/24.	24 24 24 24 24 24 24 24 24 24 24 24 24 24 (R/N each)	Y/N
Total points for Part 7 — Retest child if < 8/10.	10 10 10 10 10 10 10 10 10 10 10 10 10 10 (R/N each)	Y/N

Individually Administered Part 8–Shapes

Question	Answer	Points	(student columns)
a	6	1 or 0	
a	square	1 or 0	
b	No	1 or 0	
b	No	1 or 0	
c	No	1 or 0	
c	No	1 or 0	
d	Yes	1 or 0	
d	Yes	1 or 0	

Total points for Part 8 — Retest child if < 7/8.	8 8 8 8 8 8 8 8 8 8 8 8 8 8 (R/N each)	Y/N

Part 9–Symbol ID

Equation	Points	(student columns)
10 + 3 = 13	2 or 0	
10 + 8 = 18	2 or 0	
10 + 5 = 15	2 or 0	
10 + 9 = 19	2 or 0	
10 + 2 = 12	2 or 0	

Total points for Part 9 — Retest child if < 8/10.	10 10 10 10 10 10 10 10 10 10 10 10 10 10 (R/N each)	Y/N
Test 11 Total	100 100 100 100 100 100 100 100 100 100 100 100 100 100	
Grade		

Connecting Math Concepts

Mastery Test 12
Group Summary Sheet

Name

Group Remedy needed if > 1/4 children fail

Total points for Part 1	10	10	10	10	10	10	10	10	10	10	10	10	10	10	10	
Retest child if < 10/10.	R N	R N	R N	R N	R N	R N	R N	R N	R N	R N	R N	R N	R N	R N	R N	Y N
Total points for Part 2	12	12	12	12	12	12	12	12	12	12	12	12	12	12	12	
Retest child if < 9/12.	R N	R N	R N	R N	R N	R N	R N	R N	R N	R N	R N	R N	R N	R N	R N	Y N
Total points for Part 3	12	12	12	12	12	12	12	12	12	12	12	12	12	12	12	
Retest child if < 9/12.	R N	R N	R N	R N	R N	R N	R N	R N	R N	R N	R N	R N	R N	R N	R N	Y N
Total points for Part 4	20	20	20	20	20	20	20	20	20	20	20	20	20	20	20	
Retest child if < 16/20.	R N	R N	R N	R N	R N	R N	R N	R N	R N	R N	R N	R N	R N	R N	R N	Y N
Total points for Part 5	16	16	16	16	16	16	16	16	16	16	16	16	16	16	16	
Retest child if < 12/16.	R N	R N	R N	R N	R N	R N	R N	R N	R N	R N	R N	R N	R N	R N	R N	Y N
Total points for Part 6	18	18	18	18	18	18	18	18	18	18	18	18	18	18	18	
Retest child if < 15/18.	R N	R N	R N	R N	R N	R N	R N	R N	R N	R N	R N	R N	R N	R N	R N	Y N

Individually Administered Part 7–Tens Questions

	Answer	Points																
Fork & Spoons	Which is more?	1 or 0																
	Which is less?	1 or 0																
Dogs & Cats	More cats?	1 or 0																
	More dogs?	1 or 0																
	Numbers equal?	1 or 0																
	How many?	1 or 0																

Total points for Part 7	6	6	6	6	6	6	6	6	6	6	6	6	6	6	6	
Retest child if < 5/6.	R N	R N	R N	R N	R N	R N	R N	R N	R N	R N	R N	R N	R N	R N	R N	Y N

Part 8–Compare Groups

	What Object?	Points																
	sphere	2 or 0																
	cylinder	2 or 0																
	cone	2 or 0																

Total points for Part 8	6	6	6	6	6	6	6	6	6	6	6	6	6	6	6	
Retest child if < 6/6.	R N	R N	R N	R N	R N	R N	R N	R N	R N	R N	R N	R N	R N	R N	R N	Y N
Test 12 Total	100	100	100	100	100	100	100	100	100	100	100	100	100	100	100	
Grade																

Appendix C

Sample Lessons:

- Lesson 17 Presentation Book and Workbook
- Lesson 89 Presentation Book and Workbook

Lesson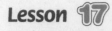

Note: (You will need a bag and 9 crayons or Board Displays
slide [17:2A] for exercise 2, steps c and d.)

EXERCISE 1: ROTE COUNTING—*Count to 15*

a. I'm going to get 10 going and count to 13. Tennn, 11, 12, 13.
• Your turn: Get 10 going. *Tennn.* Count. (Tap 3.) *11, 12, 13.*
  (Repeat step a until firm.)
b. Now I'm going to get 10 going and count to 14.
• Say 14. (Signal.) *14.*
c. Listen: Tennn, 11, 12, 13, 14. Listen again: Tennn, 11, 12, 13, 14.
• Say that part with me. *Tennn.* (Tap 4.) *11, 12, 13, 14.*
  (Repeat step c until firm.)
d. Your turn to get 10 going and count to 14. Get it going. *Tennn.*
  Count. (Tap 4.) *11, 12, 13, 14.*
  (Repeat step d until firm.)

——— INDIVIDUAL TURNS ———
(Call on individual children to get 10 going and count to 14.) *Tennn, 11,
12, 13, 14.*

EXERCISE 2: COUNTING EVENTS

a. I'll tap. You'll count each time I tap.
  (Tap) *1,* (tap) *2,* (pause) (tap) *3,* (pause) (tap) *4,* (tap) *5,* (tap) *6,* (tap)
  7, (pause) (tap) *8,* (tap) *9.*
  (Repeat until firm.)
• How many times did I tap? (Signal.) *9.*
b. I'll tap. You'll count each time I tap.
  (Tap) *1,* (pause) (tap) *2,* (tap) *3,* (pause) (tap) *4,* (pause) (tap) *5,* (tap)
  6, (tap) *7.*
  (Repeat until firm.)
• How many times did I tap? (Signal.) *7.*
c. Now I'm going to drop crayons into a bag. You'll count them.
  (Hold 9 crayons above bag.)
• Get ready. (Drop crayons.) *1, 2, 3, 4, 5, 6, 7, 8, 9.*
  (Repeat until firm.)
• How many crayons are in the bag? (Signal.) *9.*
  Yes, 9.
d. Let's see if you're right. (Quickly remove crayons and arrange them
  in a row.)
• I'll touch. You'll count. Get ready. (Touch crayons.) *1, 2, 3, 4, 5, 6,
  7, 8, 9.*
• How many crayons were in the bag? (Touch.) *9.*

$$= \quad 5 \quad 6 \quad \square \quad 7 \quad +$$

EXERCISE 3: SYMBOL IDENTIFICATION—*Introduction of 3*

a. (Display page and point to top row.)   [17:3A]
  You'll tell me the names for all the things in this row when I touch
  under them.
• (Point to =.) Get ready. (Touch.) *Equals.*
• (Point to 5.) Get ready. (Touch.) *5.*
• (Repeat for remaining symbols.) *6, box, 7, plus.*
b. Let's do those again. You'll tell me the name for each thing when I
  touch under it. I'll start at this end.
• (Point to +.) Get ready. (Touch.) *Plus.*
• (Repeat for remaining symbols.) *7, box, 6, 5, equals.*
  (Repeat steps a and b until firm.)

c. (Point to .) Here's a new number. This is 3.
• What number? (Touch.) *3.*
d. (Point to first 3.) Is this 3? (Touch.) *Yes.*
• What is it? (Touch.) *3.*
e. (Point to 5.) Is this 3? (Touch.) *No.*
• What is it? (Touch.) *5.*
f. (Repeat step e for the following:)

(Point to ___)	Is this 3?	What is it?
+	No	Plus
6	No	6
3	Yes	3
4	No	4

(Repeat symbols that were not firm.)

$$\boxed{3}$$

$$3 \quad \mathbf{5} \quad + \quad 6 \quad 3 \quad 4$$

EXERCISE 4: NEXT NUMBER

a. I'm going to count, and you're going to tell me the next number.
- Listen: 10, 11, tweeelve. What's the next number? (Signal.) *13.*
- Listen: 7, eieieight. What's the next number? (Signal.) *9.*
- Listen: 9, 10, elevennn. What's the next number? (Signal.) *12.*
- Listen: 8, niiine. What's the next number? (Signal.) *10.*
- Listen: 5, siiix. What's the next number? (Signal.) *7.*
- Listen: 6, sevennn. What's the next number? (Signal.) *8.*
- Listen: 8, 9, tennn. What's the next number? (Signal.) *11.*
- (Repeat step a until firm.)

b. Let's do it again, but this time I'm going to make it even harder.
- Listen: sevennn. What's the next number? (Signal.) *8.*
- Listen: 11, tweeelve. What's the next number? (Signal.) *13.*
- Listen: niiine. What's the next number? (Signal.) *10.*
- Listen: 10, elevennn. What's the next number? (Signal.) *12.*
- Listen: eieieight. What's the next number? (Signal.) *9.*
- Listen: 9, tennn. What's the next number? (Signal.) *11.*
- Listen: fiiive. What's the next number? (Signal.) *6.*
- Listen: fouuur. What's the next number? (Signal.) *5.*
- Listen: siiix. What's the next number? (Signal.) *7.*
- (Repeat step b until firm.)

=== INDIVIDUAL TURNS ===

(Call on individual children to perform one or two of the following tasks.)

- Listen: 7, eieieight. What's the next number? (Call on a child.) *9.*
- Listen: fiiive. What's the next number? (Call on a child.) *6.*
- Listen: 9, tennn. What's the next number? (Call on a child.) *11.*
- Listen: 11, tweeelve. What's the next number? (Call on a child.) *13.*
- Listen: 6, sevennn. What's the next number? (Call on a child.) *8.*

EXERCISE 5: LINES FOR NUMERALS

a. A number tells you how many lines to make.
- What does a number tell you? (Signal.) *How many lines to make.*
- (Repeat step a until firm.)

b. If the number is 12, how many lines do you make? (Signal.) *12.*
- If the number is zero, how many lines do you make? (Signal.) *Zero.*
- If the number is 9, how many lines do you make? (Signal.) *9.*

c. Listen: A box is not a number, so it doesn't tell how many lines to make.
- Is a box a number? (Signal.) *No.*
- Does a box tell how many lines to make? (Signal.) *No.*

d. Is 10 a number? (Signal.) *Yes.*
   Does 10 tell how many lines to make? (Signal.) *Yes.*
   How many? (Signal.) *10.*

e. Is 3 a number? (Signal.) *Yes.*
   Does 3 tell how many lines to make? (Signal.) *Yes.*
   How many? (Signal.) *3.*

f. (Write on the board:)   [17:5A]

   5 ☐ ☐

   One of these is a number, and some are boxes. If it is a number, you'll tell me what number it is.
- (Point to **5**.) What's this? (Touch.) *5.*
- Does 5 tell how many lines to make? (Touch.) *Yes.*
- How many? (Touch.) *5.*

g. (Point to first ☐.) What's this? (Touch.) *(A) box.*
- Does a box tell how many lines to make? (Touch.) *No.*

h. (Point to next ☐.) What's this? (Touch.) *(A) box.*
- Does a box tell how many lines to make? (Touch.) *No.*

i. (Point to **5**.) I'm going to make lines for this number.
- How many lines? (Touch.) *5.*
- Count the lines and tell me when to stop. Get ready. (Make lines under 5.) *1, 2, 3, 4, 5, stop.*
   (Teacher reference:)   [17:5B1–5]

   5 ☐ ☐
   |||||

   We made the lines for the number 5.

j. (Point to boxes.) What are these? (Touch.) *Boxes.*
- Do they tell how many lines to make? (Touch.) *No.*
- So I don't make lines for them.

7 5 3

EXERCISE 6: SYMBOL IDENTIFICATION

a. (Display page.)   [17:6A]
   One of these numbers is 3.
- (Point to **3.**) Here's the new number. What number? (Touch.) *3.*

b. (Point to **7.**) Is this 3? (Touch.) *No.*
- What is it? (Touch.) *7.*
   Yes, 7.
- (Point to **5.**) Is this 3? (Touch.) *No.*
- What is it? (Touch.) *5.*
   Yes, 5.
- (Point to **3.**) Is this 3? (Touch.) *Yes.*

c. You'll tell me the names again when I touch under each thing.
- (Point to **7.**) Get ready. (Touch.) *7.*
- (Point to **5.**) Get ready. (Touch.) *5.*
- (Point to **3.**) Get ready. (Touch.) *3.*
   (Repeat step c until firm.)

EXERCISE 7: COUNTING FROM NUMBERS

a. I'm going to get numbers going and say the next number.
   My turn to get 6 going and say the next number. Siiix. (Signal.) 7.
- My turn to get 4 going and say the next number. Fouuur. (Signal.) 5.
- My turn to get 7 going and say the next number. Sevennn.
   (Signal.) 8.
- Your turn to get 7 going and say the next number. *Sevennn.*
   (Signal.) *8.*
- Get 4 going. *Fouuur.* Next number. (Signal.) *5.*
- Get 6 going. *Siiix.* Next number. (Signal.) *7.*
   (Repeat step a until firm.)

b. This time we're going to get numbers going and count.
- My turn to get 8 going and count. Eieieight, 9, 10, 11.
- Your turn to get it going and count. Get 8 going. *Eieieight.* Count.
   (Tap 3.) *9, 10, 11.*

c. My turn to get 5 going and count. Fiiive, 6, 7, 8.
- Your turn to get it going and count. Get 5 going. *Fiiive.* Count.
   (Tap 3.) *6, 7, 8.*

d. Get 7 going. *Sevennn.* Count. (Tap 3.) *8, 9, 10.*
- Get 2 going. *Twooo.* Count. (Tap 3.) *3, 4, 5.*
   (Repeat step d until firm.)

═══ INDIVIDUAL TURNS ═══
(Call on individual children to perform one of the following tasks.)

- Get 7 going and count. *Sevennn.* (Tap 3.) *8, 9, 10.*
- Get 2 going and count. *Twooo.* (Tap 3.) *3, 4, 5.*

EXERCISE 8: PLUS

a. (Display page and point to **+**.)   [17:8A]
- Everybody, what is this? (Touch.) *Plus.*
   Yes, plus.

b. When you plus, you get more. What do you do when you plus?
   (Signal.) *(You) get more.*

c. If you plus 1, you get 1 more.
- What happens if you plus 1? (Signal.) *(You) get 1 more.*
- What happens if you plus 5? (Signal.) *(You) get 5 more.*
- What happens if you plus 15? (Signal.) *(You) get 15 more.*
- What happens if you plus 7? (Signal.) *(You) get 7 more.*
   (Repeat step c until firm.)

7

EXERCISE 9: COUNTING TWO GROUPS

a. (Display page and point to lines.)   [17:9A]
  Here are two groups of lines. You're going to count the lines in each group.
• (Point to IIIIIII.) Count the lines in this group. Get ready.
  (Touch lines.) *1, 2, 3, 4, 5, 6, 7.*
• How many lines are in this group? (Touch.) *7.*
b. (Point to II.) Count the lines in this group. Get ready.
  (Touch lines.) *1, 2.*
• How many lines are in this group? (Touch.) *2.*
c. (Point to IIIIIII.) I'm going to count the lines in both groups.
  (Touch lines in first group.) 1, 2, 3, 4, 5, 6, sevennn.
  (Touch lines in second group.) 8, 9.
• This time I'll count the first group. Then you'll keep on counting.
  (Touch lines in first group.) 1, 2, 3, 4, 5, 6, sevennn.
  (Touch lines in second group.) *8, 9.*
  (Repeat until firm.)
• How many lines in both groups? (Signal.) *9.*

d. You'll do the hard part again. (Touch line 7.) This is line 7.
• Which line? (Touch.) *7.*
  You'll get it going and count the rest of the lines.
• Get 7 going. *Sevennn.* (Touch lines in second group.) *8, 9.*
e. Let's do the hard part again.
  I'll touch line 7. (Touch.)
• Get 7 going. *Sevennn.* (Touch lines in second group.) *8, 9.*
  (Repeat until firm.)
  How many lines are in both groups? (Signal.) *9.*

EXERCISE 10: MAKING LINES

a. (Write on the board:)   [17:10A]

  I'm going to count to 3 and make a line for each number I count. For each number, I start at the big ball and make a line down. Watch.
• (Put chalk on first big ball.) One. (Make line and put chalk on next big ball.) Two. (Make line and put chalk on next big ball.) Three. (Make line.)
• (Point to lines.) How many lines did I make? (Touch.) *3.*
• After I made each line, did you see how quickly I got ready to make the next line? (Children respond.)
b. (Open workbooks to Lesson 17 and distribute to children.)
• Now **you're** going to make a line for each number I count. (Hold up worksheet and point to snake.)

c. Touch the snake on your worksheet. ✔
• Touch the big ball next to the snake. ✔
  That's where you'll start your first line.
• After I say 1, you'll make the line for that ball. What will you do after I say 1? (Signal.) *Make the line for that ball.*
• Will you make the line for that ball before I say 1? (Signal.) *No.*
  That's right. After I say 1, you'll start at the big ball and make the line down.
d. Touch the ball where you'll start your next line. ✔
• After I say 2, you'll make the line for that ball. What will you do after I say 2? (Signal.) *Make the line for that ball.*
• Will you make the line for that ball before I say 2? (Signal.) *No.*
  That's right. After I say 2, you'll start at the big ball and make the line down.
e. Touch the ball where you'll start your next line. ✔
• After I say 3, you'll make the line for that ball. What will you do after I say 3? (Signal.) *Make the line for that ball.*
• Will you make the line for that ball before I say 3? (Signal.) *No.*
  That's right. After I say 3, you'll start at the big ball and make the line down.
  (Repeat steps c through e until firm.)
f. Now put your pencil on the ball where you'll start your first line. ✔
• I'm going to count to 3. You'll make a line after I say each number. Get ready. One. (Children make a line.) Two. (Children make a line.) Three. (Children make a line.)
g. (If children did not make mistakes on step f, go to Exercise 11.)
• (If children made mistakes on step f, repeat steps c through f with the dotted lines next to the **spoon**.)

EXERCISE 11: COUNTING LINES

a. Touch the butterfly on your worksheet. ✔
(Teacher reference:)

You're going to count the group of lines next to the butterfly.
- Put your finger over the first line. ✔
- Touch and count the lines. Get ready. (Tap 6.) *1, 2, 3, 4, 5, 6.*
- (Repeat until firm.)
- How many lines in that group? (Signal.) *6.*
b. Touch the car. ✔
You're going to count the group of lines next to the car.
- Put your finger over the first line. ✔
- Touch and count the lines. Get ready. (Tap 5.) *1, 2, 3, 4, 5.*
- (Repeat until firm.)
- How many lines in that group? (Signal.) *5.*

━━━━━ INDIVIDUAL TURNS ━━━━━

(Call on individual children to perform one of the following tasks.)

- Touch and count the group of lines next to the butterfly. *1, 2, 3, 4, 5, 6.*
  How many lines? *6.*
- Touch and count the group of lines next to the car. *1, 2, 3, 4, 5.*
  How many lines? *5.*

EXERCISE 12: SYMBOL WRITING

a. Touch the alligator on your worksheet. ✔
(Teacher reference:)

What's the number next to the alligator? (Signal.) *5.*
- Complete the row of 5s.
  (Observe children and give feedback.)
b. Touch the duck. ✔
- Touch what's next to the duck. ✔
- Everybody, what are you touching? (Signal.) *Equals.*
- What's the number next to the equals? (Signal.) *2.*
- What's next to the 2? (Signal.) *4.*
  Yes, 4.
c. Touch the dog. ✔
- Everybody, what's next to the dog? (Signal.) *7.*
- What's next to the 7? (Signal.) *6.*
  Yes, 6.
d. Complete the rows.
  (Observe children and give feedback.)

EXERCISE 13: CROSS-OUT/CIRCLE GAME

a. Touch the cross-out/circle game on your worksheet. ✔
(Teacher reference:)

- You are going to cross out some things and circle other things.
b. Touch the thing that's crossed out. ✔
- What are you going to cross out? (Signal.) *2.*
c. Touch the thing that's circled. ✔
- What are you going to circle? (Signal.) *7.*
  (Repeat steps b and c until firm.)
d. Again, what are you going to cross out? (Signal.) *2.*
- Cross out all the 2s.
  (Observe children and give feedback.)
e. You crossed out the 2s. Now you have to do something else.
- What do you have to do now? (Call on a child. Idea: *Circle all the 7s.*)
  You have to circle all the 7s. Do it.
  (Observe children and give feedback.)

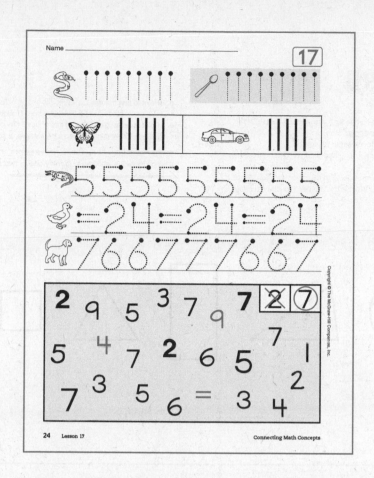

24   Lesson 17

Connecting Math Concepts

Lesson

EXERCISE 1: SHAPES

a. You learned another name for a box. What's another name for a box? (Signal.) *A rectangle.*

b. (Display page and touch either triangle.)   [89:1A]
This shape has three sides. It's a triangle. What is this shape? (Touch.) *A triangle.*

c. (Point to objects.) You're going to tell me what these things are.

• (Point to first circle.) What is this shape? (Touch.) *(A) circle.*

d. (Point to first triangle.) What is this shape? (Touch.) *(A) triangle.*

• How many sides does a triangle have? (Touch.) *Three.*

e. (Point to rectangle.) What is this shape? (Touch.) *A rectangle.*

• (Repeat for remaining objects.) *A triangle, a square, a circle.*

━━━━ INDIVIDUAL TURNS ━━━━
(Call on individual children to identify one or two shapes.)

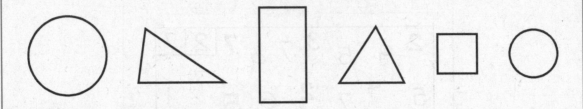

$$1 + 16 =$$

$$5 + 80 = \qquad \mid \qquad 1 + 7 =$$

EXERCISE 2: TURN-AROUND EQUATIONS

a. (Display page and point to problems.)   [89:2A]
You don't know the answer to these problems, but you can figure out the answers by saying the turn-around equations.

• (Point to **1 + 16.**) Tell me the turn-around for 1 plus 16. Get ready. (Signal.) *16 plus 1.*

• What does 16 plus 1 equal? (Signal.) *17.*
So what does 1 plus 16 equal? (Signal.) *17.*

• Say the equation for 1 plus 16. (Signal.) *1 plus 16 equals 17.*
(Repeat until firm.)

b. (Point to **5 + 80.**) Tell me the turn-around for 5 plus 80. Get ready. (Signal.) *80 plus 5.*

• What does 80 plus 5 equal? (Signal.) *85.*
So what does 5 plus 80 equal? (Signal.) *85.*

• Say the equation for 5 plus 80. (Signal.) *5 plus 80 equals 85.*

c. (Point to **1 + 7.**) Tell me the turn-around for 1 plus 7. Get ready. (Signal.) *7 plus 1.*

• What does 7 plus 1 equal? (Signal.) *8.*
So what does 1 plus 7 equal? (Signal.) *8.*

• Say the equation for 1 plus 7. (Signal.) *1 plus 7 equals 8.*
(Repeat steps b and c that were not firm.)

EXERCISE 3: COINS—*Identifying Quarters*

a. (Display page and point to quarter in box.)   [89:3A]
- This coin is a **quarter.** What kind of coin is this? (Touch.) *A quarter.*
- (Point to first row.) I'll point to coins. You'll tell me what they are when I touch them.
- (Point to first quarter.) What is this? (Touch.) *(A) quarter.*
- (Point to first dime.) What is this? (Touch.) *(A) dime.*
- (Repeat for remaining coins.) *A penny, a nickel, a quarter, a penny, a dime, a quarter.*
- (Repeat until firm.)

---

1 2 3 4 5

EXERCISE 4: COUNTING BACKWARD

a. (Display page and point to number line.)   [89:4A]
    You're going to count backward on this number line.
- (Point to **5.**) Count backward with me. Get ready. (Touch symbols.) *5, 4, 3, 2, 1.*
- (Repeat until firm.)

b. All by yourself. I'll touch the numbers. You count backward.
- (Point to **5.**) Get ready. (Touch symbols.) *5, 4, 3, 2, 1.*
- (Repeat step b until firm.)

c. (Stop displaying page.) I can count backward from 3 without looking. Listen: 3, 2, 1.
- Count backward from 3 with me. Get ready. (Count with children:) *3, 2, 1.*
- All by yourself. Count backward from 3. Get ready. (Tap 3.) *3, 2, 1.*
- (Repeat until firm.)

—————— INDIVIDUAL TURNS ——————
(Call on individual children to count backward from 3.)

134   103   150
172   113

EXERCISE 5: SYMBOL IDENTIFICATION—
Three-Part Numbers

a. (Display page and point to numbers.) [89:5A]
These three-part numbers don't have an underlined part. Remember, you say **one hundred** for the 1.
• What do you say for the 1? (Signal.) *One hundred.*
Then you read the rest of the number.

b. (Point to **134.**) What do you say for the 1? (Touch.) *One hundred.*
• What do you say for the other part? (Touch.) *34.*
• Say the whole number. (Touch.) *One hundred 34.*

c. (Point to **103.**) What do you say for the 1? (Touch.) *One hundred.*
• What do you say for the other part? (Touch.) *3.*
• Say the whole number. (Touch.) *One hundred 3.*

d. For the rest of the numbers, you'll read the whole number.
• (Point to **150.**) What's the whole number? (Touch.) *One hundred 50.*
  (To correct:)___
  • (Point to **150.**) What do you say for 1? (Touch.) *One hundred.*
  • What do you say for the other part? (Touch.) *50.*
  • Say the whole number. (Touch.) *One hundred 50.*
• (Point to **172.**) What number? (Touch.) *172.*
• (Point to **113.**) What number? (Touch.) *113.*
  (Repeat steps b through d until firm.)

EXERCISE 6: ORDINAL NUMBERS

a. (Display page and point to dog.) [89:6A]
• Count the things in this row. Get ready. (Touch.) *1, 2, 3, 4, 5.*
• How many things are in this row? (Signal.) *5.*
• (Point to dog.) This is the first thing. What is the first thing in this row? (Touch.) *(A) dog.*
• Say **first.** (Signal.) *First.*

b. (Point to cat.) This is the **second** thing. What is the second thing in this row? (Touch.) *(A) cat.*
• Say **second.** (Signal.) *Second.*

c. (Point to boy.) This is the **third** thing. What is the third thing in this row? (Touch.) *(A) boy.*
• Say **third.** (Signal.) *Third.*

d. (Point to dog.) My turn: (Touch dog.) First. (Touch cat.) Second. (Touch boy.) Third.
• Say that with me. (Touch objects.) *First, second, third.*
  (Repeat step d until firm.)

e. What's the first thing in the row? (Signal.) *(A) dog.*
• What's the second thing in the row? (Signal.) *(A) cat.*
• What's the third thing in the row? (Signal.) *(A) boy.*
  (Repeat step e until firm.)

EXERCISE 7: ADDING TWO-PART NUMBERS

a. Start with 20 and count by tens to one hundred. Get 20 going. *Twentyyy.* Count. (Tap 8.) *30, 40, 50, 60, 70, 80, 90, one hundred.* (Repeat step a until firm.)

b. Now you'll start with 29 and plus tens to 99. What number will you start with? (Signal.) *29.*
- Get 29 going. *Twenty-niiine.* Plus tens. (Tap 7.) *39, 49, 59, 69, 79, 89, 99.*
- Now you'll start with 59 and plus **ones** to 64. Get 59 going. *Fifty-niiine.* Plus ones. (Tap 5.) *60, 61, 62, 63, 64.*
- Now you'll start with 43 and plus **ones** to 50. Get 43 going. *Forty-threee.* Plus ones. (Tap 7.) *44, 45, 46, 47, 48, 49, 50.*
- Now start with 33 and plus **tens** to 93. Get 33 going. *Thirty-threee.* Plus tens. (Tap 6.) *43, 53, 63, 73, 83, 93.*
(Repeat step b until firm.)

c. Today's lesson is 89. What number? (Signal.) *89.*
- Tell me the parts of 89. Get ready. (Signal.) *8 and 9.*
- (Distribute unopened workbooks to children.) Open your workbook to Lesson 89.
(Observe children and give feedback.)

d. Touch the problem 59 plus 40 ☞ equals. ✔
You'll make the counters for one of the numbers.
- Which number will you make counters for? (Signal.) *40.*
- Will you make Ts or lines? (Signal.) *Ts.*
- How many Ts will you make for 40? (Signal.) *4.*

- So you'll make 4 Ts. What will you make for the problem 59 plus 40? (Signal.) *4 Ts.*
e. You'll tell me what you'll make for the rest of the problems.
- Touch 59 plus 4. Think about how many Ts or lines you'll make. (Pause.)
- What will you make for the problem 59 plus 4? (Signal.) *4 lines.* (Repeat until firm.)
f. Touch 43 plus 30. ✔
- What will you make for the problem 43 plus 30? (Signal.) *3 Ts.*
g. Make the Ts or lines for each problem.
(Observe children and give feedback.)
(Teacher reference:)

```
59+40=     59+4=
 T T T T    / / / /
   43+30=
    T T T
```

h. Touch the problem 59 plus 40 equals again. ✔
You'll get the number going and touch and count for the Ts to figure out the answer. Remember how many you plus for each T.
- Touch the number you'll get going. ✔
- Get it going. *Fifty-niiine.* Count for the Ts. (Tap 4.) *69, 79, 89, 99.* (Repeat until firm.)
- What's the answer? (Signal.) *99.*
i. Touch the problem 59 plus 4. ✔
You'll get a number going and touch and count for the lines to figure out the answer.
- Touch the number you'll get going. ✔
- Get it going. *Fifty-niiine.* Count for the lines. (Tap 4.) *60, 61, 62, 63.* (Repeat until firm.)
- What's the answer? (Signal.) *63.*

---

j. Touch the problem 43 plus 30. ✔
You'll get the number going and touch and count for the Ts to figure out the answer.
- Touch the number you'll get going. ✔
- Get it going. *Forty-threee.* Count for the Ts. (Tap 3.) *53, 63, 73.* (Repeat until firm.)
- What's the answer? (Signal.) *73.*
k. Go back to the first problem. Touch and count to yourself for both groups and write the answer. Then work the other problems. Remember to plus one for lines and plus ten for Ts. Put your pencil down when you've completed all the equations.
(Observe children and give feedback.)
(Answer key:)

```
59+40=99   59+4= 63
 T T T T    / / / /
    43+30= 73
     T T T
```

l. Check your work.
- Touch and read the first equation. Get ready. (Tap 5.) *59 plus 40 equals 99.*
- Touch and read the next equation. Get ready. (Tap 5.) *59 plus 4 equals 63.*
- Touch and read the last equation. Get ready. (Tap 5.) *43 plus 30 equals 73.*

EXERCISE 8: COLUMN PROBLEMS

a. Find the column problem 70 plus 6 equals 76 on worksheet 89. ✔
(Teacher reference:)

```
 70   42   29   50   28
+ 6  - 0  + 1  + 3  + 0
 76   42   30   53   28
```

The answers to these problems are right, but some of the answers are in the wrong place.
- Touch 70 plus 6. ✔
- What's the answer? (Signal.) *76.*
- Is 76 written in the right place? (Signal.) *No.*
b. Touch 42 take away zero. ✔
- What's the answer? (Signal.) *42.*
- Is the answer written in the right place? (Signal.) *Yes.*
c. Touch 29 plus 1. ✔
- What's the answer? (Signal.) *30.*
- Is 30 written in the right place? (Signal.) *No.*
d. Touch 50 plus 3. ✔
- What's the answer? (Signal.) *53.*
- Is 53 written in the right place? (Signal.) *Yes.*
e. Touch 28 plus zero. ✔
- What's the answer? (Signal.) *28.*
- Is 28 written in the right place? (Signal.) *Yes.*
f. Your turn: Cross out the answers that are written in the wrong place. Make one line through those answers.
(Observe children and give feedback.)
(Answer key:)

```
 70   42   29   50   28
+ 6  - 0  + 1  + 3  + 0
 76   42   30   53   28
```

a. I'll say two numbers. You'll say them; then you tell me which is more.
- Listen: 180 and one hundred 8. Say 180 and 108. (Signal.) *180 and 108.*
- Which is more? (Signal.) *180.*
- Listen: 21 and 12. Say 21 and 12. (Signal.) *21 and 12.*
- Which is more? (Signal.) *21.*
- (Repeat step a until firm.)

b. I'll say two weights. You'll say them; then you tell me which is heavier.
- Listen: 63 ounces and 47 ounces. Say 63 ounces and 47 ounces. (Signal.) *63 ounces and 47 ounces.*
- Which is heavier? (Signal.) *63 ounces.*
- Listen: 29 pounds and 38 pounds. Say 29 pounds and 38 pounds. (Signal.) *29 pounds and 38 pounds.*
- Which is heavier? (Signal.) *38 pounds.*
- Listen: 17 pounds and 15 pounds. Say 17 pounds and 15 pounds. (Signal.) *17 pounds and 15 pounds.*
- Which is heavier? (Signal.) *17 pounds.*
- Listen: 19 tons and 21 tons. Say 19 tons and 21 tons. (Signal.) *19 tons and 21 tons.*
- Which is heavier? (Signal.) *21 tons.*
- (Repeat step b until firm.)

c. Touch the numbers 17 and 15 on worksheet 89. ✔
(Teacher reference:)

17 15 | 12 21 | 180 108

You're going to circle the numbers that are more.
- Which is more, 17 or 15? (Signal.) *17.*
d. Touch the next group of numbers. ✔
- What's the first number? (Signal.) *12.*
- What's the other number? (Signal.) *21.*
- Which is more, 12 or 21? (Signal.) *21.*
e. Touch the last group of numbers. ✔
- What's the first number? (Signal.) *180.*
- What's the other number? (Signal.) *108.*
- Which is more, 180 or 108? (Signal.) *180.*
f. Circle the number that is more in each group.
(Observe children and give feedback.)
(Answer key:)

⟨17⟩ 15 | 12 ⟨21⟩ | ⟨180⟩ 108

---

a. Touch the problem 65 plus how many equals 70 on worksheet 89. ✔
(Teacher reference:)

$$65+\square=70 \quad | \quad 8-3=\square$$
$$5+4=\square \quad | \quad 19+\square=23$$

Some of these problems plus, and one takes away. For some of the plus problems, you'll count and make lines to figure out the number you plus.

b. Touch and read the first problem. Get ready. (Tap 5.) *65 plus how many equals 70.*
- To work that problem, will you make lines for a number? (Signal.) *No.*
- What does that problem tell you to do? (Signal.) *Start with 65 and count to 70.*
c. Touch and read the next problem. Get ready. (Tap 5.) *8 take away 3 equals how many.*
- To work that problem, will you make lines for a number? (Signal.) *Yes.*
- What number will you make lines for? (Signal.) *8.*
d. Touch and read the next problem. Get ready. (Tap 5.) *5 plus 4 equals how many.*
- To work that problem, will you make lines for a number? (Signal.) *Yes.*
- What number will you make lines for? (Signal.) *4.*
e. Touch and read the last problem. Get ready. (Tap 5.) *19 plus how many equals 23.*
- To work that problem, will you make lines for a number? (Signal.) *No.*
- What does that problem tell you to do? (Signal.) *Start with 19 and count to 23.*
- (Repeat steps b through e until firm.)
You'll work these problems later.

Coins

a. Turn to the other side of worksheet 89 and touch the first group of coins. ✔
You'll count the cents for each group.
- Touch the first coin. ✔
- What coin are you touching? (Signal.) *(A) nickel.*
- What number do you get going for a nickel? (Signal.) *5.*
- Get it going. *Fiiive.* Count for the pennies. (Tap 5.) *6, 7, 8, 9, 10.*
- (Repeat until firm.)
- How many cents are in that group? (Signal.) *10.*
b. Touch the next group of coins. ✔
- Touch the first coin. ✔
- What coin are you touching? (Signal.) *(A) dime.*
- What do you count by for each dime? (Signal.) *10.*
- What's the other kind of coin in that group? (Signal.) *Penny.*
- Finger over the first dime. ✔
- Touch and count for the dimes. Get ready. (Tap 5.) *10, 20, 30, 40, fiftyyy.* Count for the pennies. (Tap 4.) *51, 52, 53, 54.*
- (Repeat until firm.)
- How many cents are in that group? (Signal.) *54.*
Later you'll count the cents for each group again and write an equals sign and the number.

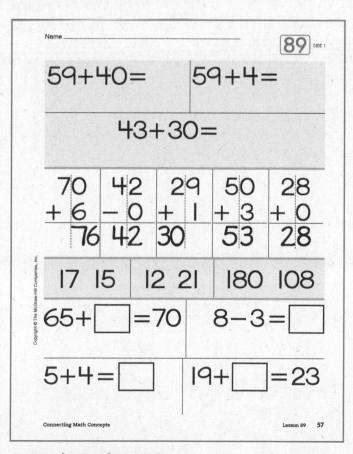

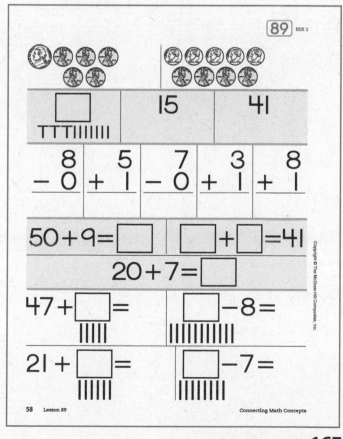